KNOWLEDGE THROUGH COLOR

STARS, PLANETS, AND GALAXIES

BY SUNE ENGELBREKTSON

A RIDGE PRESS BOOK/BANTAM BOOKS
TORONTO   NEW YORK   LONDON

Photo Credits

STARS, PLANETS, AND GALAXIES
A Bantam Book published by arrangement with The Ridge Press, Inc.
Text prepared under the supervision of Laurence Urdang Inc.
Designed and produced by The Ridge Press, Inc. All rights reserved.
Copyright 1975 in all countries of the International Copyright Union
by The Ridge Press, Inc. This book may not be reproduced in whole or in
part by mimeograph or any other means, without permission. For
information address: The Ridge Press, Inc., 25 West 43rd Street,
New York, N.Y. 10036.

Library of Congress Catalog Card Number: 75-516
Published simultaneously in the United States and Canada.

Bantam Books are published by Bantam Books, Inc.
Its trademark, consisting of the words "Bantam Books" and the portrayal
of a bantam, is registered in the United States Patent Office
and in other countries. Marca Registrada.

Bantam Books, Inc., 666 Fifth Avenue, New York, N.Y. 10019.
Printed in Italy by Mondadori Editore, Verona.

Contents

Part 1•The Visible Sky

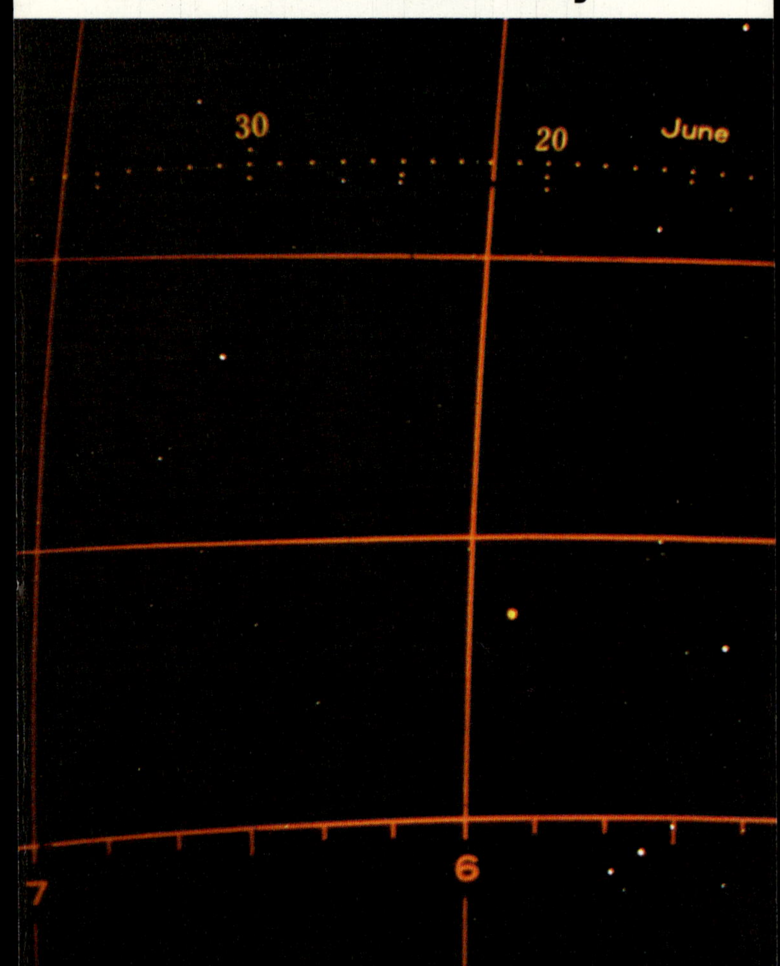

Orion region with celestial
coordinates and the ecliptic
marking the annual
path of the sun

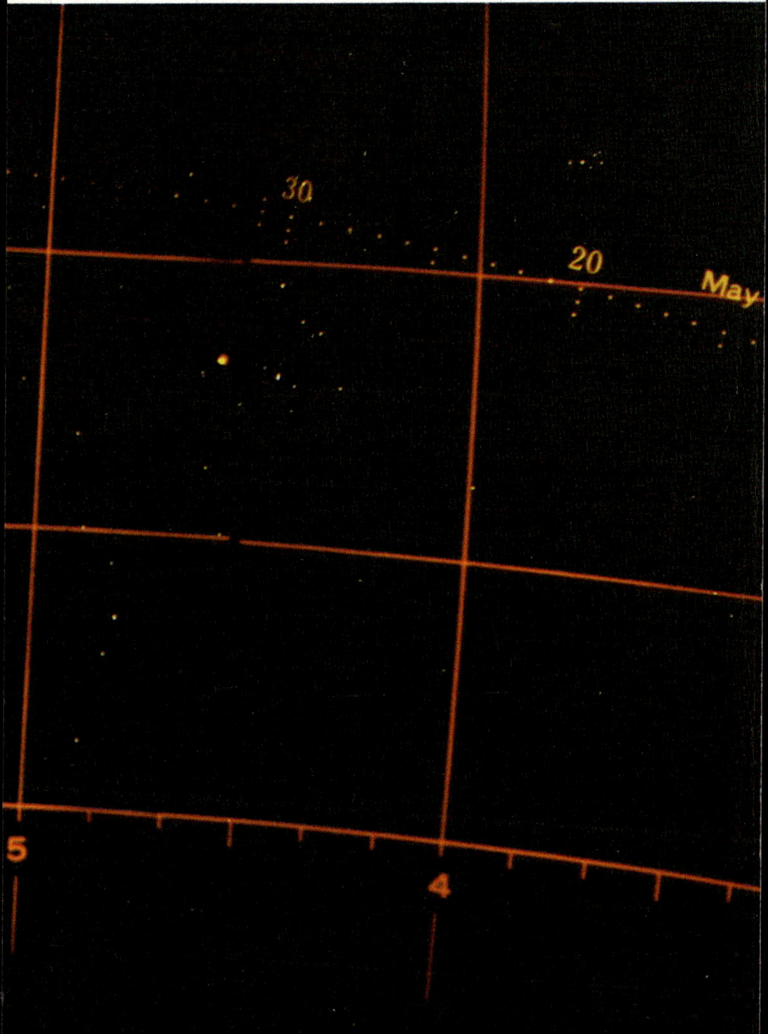

30
20
May
5
4

The Meaning of Astronomy

From the very beginning, man has been fascinated by the universe around him. To the early observer, it seemed as though he stood at the center of all things, with the ability to seek understanding of himself and of the role he played in the cosmic scene. The star-studded heavens inspired his philosophies.

There was fear of the unknown—forces at work that defied explanation. The celestial objects had meaningful movements seeking interpretation and man had to know if these signs were meant for him.

The sun and the moon are the most conspicuous objects in the sky, and it was only natural to assume that they ruled the day and the night. During eclipses, the sun and moon caused consternation. At the time of a solar eclipse, the midday sky darkens as the moon moves across the face of the sun. When the moon is eclipsed, it passes into the earth's shadow and is immersed in the reddish light of twilight. There is evidence that as early as neolithic times, man was already capable of predicting these frightening events.

Occasionally, a comet would appear and be greeted as an evil sign. These omens of bad tidings were called *aster kometes* (``long-haired stars'') by the Greeks. The sun, the moon, and the five planets visible to the unaided eye were called *planetes* (``wanderers'') by the Greeks. Today the word *planet* refers to the bodies, including the earth, that revolve about the sun. The sun is a star, but the moon is the natural satellite of the earth. The movements of the seven ancient *planetes* were believed to affect the destiny of man, and this belief that cosmic reason gives order to the universe was the beginning of *astrology*.

But the celestial objects were also studied for more practical purposes such as keeping time and measuring the days, the seasons, and the year. As knowledge increased, the superstitious beliefs of astrology gave way to more objective investigations and *Astronomy*, the science of the heavens, which began with man's fear of the unknown, continues his search for a better understanding of the universe.

Mythology of the Heavens

After sunset, the first stars and sometimes a planet or two appear in the twilight glow. In the absence of the moon, the sky grows dark and the faintest stars come into view. The stars seem attached to the vault of the heavens all at the same distance from the earth. To the ancients, stars clustered in the same area were thought to be near each other, forming various patterns called *constellations* that were imagined to be the outlines of various real or fictitious creatures. Bears, lions, and serpents were visualized among the stars. In addition to animals found on earth

there were celestial unicorns, dragons, centaurs, and the demons of folklore. Other parts of the sky contained the heroes, gods, and goddesses of ancient mythology. The sky provided a means to illustrate stories that still fascinate the reader.

Each isolated civilization has found its own stories in the stars, tales that reflected the way of life of peoples with little or no mutual contact. In the Mediterranean world, the hazy band of glowing stars is called the Milky Way. To American Indians, it was the campfires of fallen braves on the way to the Happy Hunting Ground. Where interaction was possible, star names, stories, and constellations were similar. Interestingly, Ursa Major, the Great Bear, looked like a bear to the Mediterranean peoples as well as to the American Indians.

Primitive man interpreted natural
celestial phenomena, such as meteor showers,
as fearful events affecting his destiny.

Astronomy of Antiquity

Before the present era, the Greeks made notable strides in describing celestial phenomena. Eudoxus (408–355 B.C.) suggested that the heavenly bodies were attached to transparent spheres turning on separate axes. Later, Eratosthenes (276–195 B.C.) measured the circumference of the earth. He noted that on June 21, the rays of the noon sun at Syene, Egypt were reflected from a well. Syene, near modern Aswan, is at the Tropic of Cancer where the sun is at the *zenith* (directly overhead) on the June solstice, the first day of summer. To the north, at Alexandria, the sun is 7° from the zenith on that date. Since 7° is about 1/51 part of a circle, the distance between Syene and Alexandria is equal to 1/51st of the circumference of the earth. The earth was thought to be a sphere.

Hipparchus (about 150 B.C.), among many achievements, devised the concept of celestial bodies moving on epicycles and deferents. He provided Claudius Ptolemy (100–178 A.D.) with systematic observations that were summarized in the *Almagest,* the greatest astronomical work of antiquity. In the *Almagest,* Ptolemy describes his *geocentric world system* with the sun, moon, and planets moving around a central earth. The planets revolve in an epicycle, the center of which traces a deferent around the earth. The system was used with modifications long after the 16th century, when Copernicus introduced the *heliocentric hypothesis.*

From Copernicus to Newton

The 16th century was a time of great change in man's concept of his planet and its location in the cosmic system. In the age of exploration, the earth was circumnavigated, proving once and for all that the Mediterranean region could not be at the center of the world. Accepting that meant accepting the possibility that the earth was not at the center of the universe.

Nicolas Copernicus (1473–1543) described a system with the sun at the center. His heliocentric hypothesis, found in his book *De Revolutionibus,* could account for the motions of the planets without resorting to the numerous epicycles necessary for the geocentric system.

Later, Galileo Galilei (1564–1642) gave strong support to the heliocentric system with his telescopic observations. He recognized mountains on the moon, as well as maria and craters. But his discovery of the four satellites of Jupiter, called the Galilean moons, proved that other planets were capable of attracting celestial bodies. He observed the planet Venus passing through moon-like phases during its period of revolution.

About the same time, Johannes Kepler (1571–1630), using the precise tables of planetary motion of Tycho Brahe (1546–1601), formulated three laws of planetary motion: that planets revolve in elliptical orbits;

Top: Ptolemy's geocentric system described planetary motions. Btm.: Copernicus' heliocentric system, with orbits around the sun

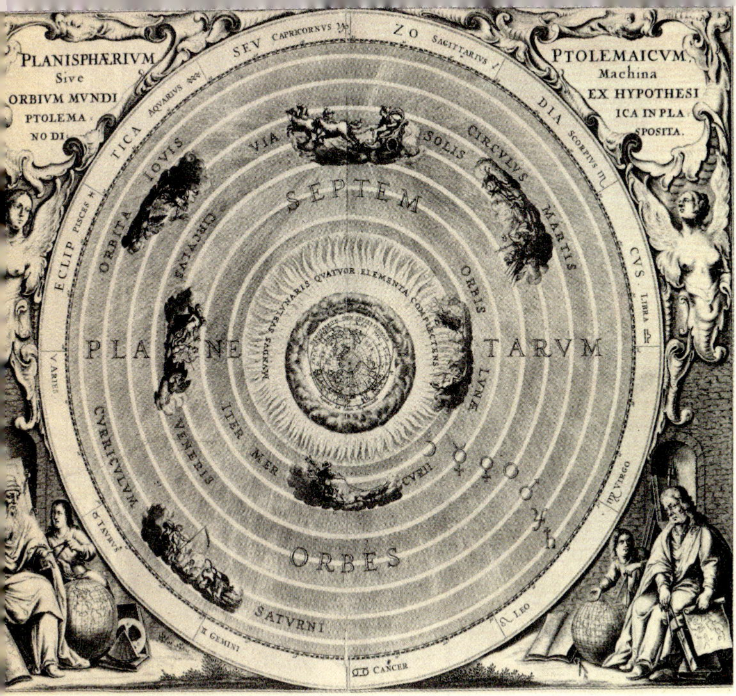

PLANISPHÆRIVM
Sive
ORBIVM MVNDI
PTOLEMA·
NO DI·

PTOLEMAICVM·
Machina
EX HYPOTHESI
ICA IN PLA·
SPOSITA.

SEPTEM
PLANETARVM
ORBES

MVNDVS SVBLVNARIS QVATVOR ELEMENTA COMPLECTENS

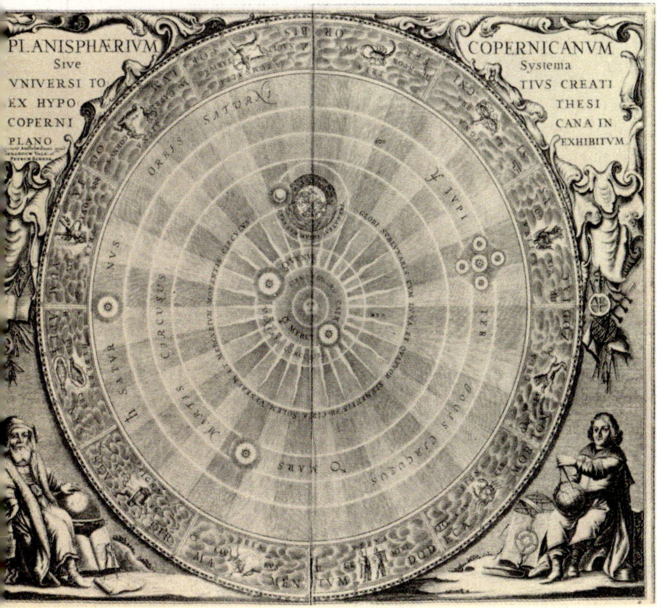

CIRCVLVS SOLIS
ORBIS LVNÆ
VIA SOLIS
ITER MER CVRII
ITER VENERIS
CVRRICVLVM
ORBITA IOVIS
CIRCVLVS MARTIS
ECLIP TICA
SATVRNI

ZODIACVS
PISCES · AQVARIVS · CAPRICORNVS · SAGITTARIVS · SCORPIVS · LIBRA · VIRGO · LEO · CANCER · GEMINI · TAVRVS · ARIES

PLANISPHÆRIVM
Sive
VNIVERSI TO
EX HYPO
COPERNI
PLANO

COPERNICANVM
Systema
TIVS CREATI
THESI
CANA IN
EXHIBITVM

ORBIS SATVRNI
ORBIS IOVIS
ORBIS MARTIS
CIRCVLVS TERRÆ
MERCVRIVS

9

that the line segment between the planet and the sun sweeps out equal areas in equal intervals of time; and that the squares of the sidereal periods of the planets are in direct proportion to the cubes of the distance between them and the sun.

Isaac Newton (1642–1727) described the motions of the planets in terms of mass, momentum, and force. To keep in an orbit around the sun, the planet accelerates toward the center. This centripetal force is gravitation. Newton expressed mathematically his law of universal gravitation, which states that every particle in the universe is attracted to every other particle with a force that is directly proportional to the product of their masses and inversely proportional to the square of their distances from each other.

Astronomy in the 19th Century
The 19th century was noted for its advances in stellar astronomy. Herschel (1738–1822), who discovered the planet Uranus in 1781, located 2,500 nebulae and 806 double stars, and, to determine stellar distribution, he counted the number of stars in designated fields. He

Isaac Newton discovers that sunlight is a mixture of a rainbow of colors called the spectrum.

called this method star *gauging*. Herschel, the founder of stellar astronomy, also made a systematic study of the apparent brightness of the stars.

The first minor planet or *asteroid,* called Ceres, was discovered by Piazzi in 1801, and the discovery of many others followed. Today, thousands of asteroids are known. A disturbance in the motion of the planet Uranus led to the discovery of Neptune in 1846. Leverrier of France and Adams of England independently located the planet mathematically—a triumph for Newton's gravitational theory. During the 19th century, many advances were made in the design of telescopes and other instruments. The invention of the spectroscope led to the discovery of *Fraunhofer lines* in solar and stellar spectra, revealing chemical compositions as well as motions in the line of sight. The camera ushered in 20th century astronomy, for it was photography that aided the discovery of the nature of nebulae and of stars made visible by long exposures.

Twentieth Century Astronomy

In 1922, using the 100-inch Hooker Telescope at Mount Wilson, Edwin Hubble photographed variable stars in the Andromeda Galaxy. These stars provided the way to measure distances to remote star systems. About 30 years later, the 200-inch Hale Telescope found galaxies by the millions extending to thousands of millions of light-years without a decrease in population. When light from the galaxies was examined, there was evidence that the galaxies were receding from each other, indicating that the universe is expanding. The discovery of nuclear energy and the relativity theories of Albert Einstein led to an understanding of the internal processes required to keep the sun and stars shining for thousands of millions of years.

In 1931, Karl Jansky designed the first radio telescope, allowing long wavelengths of radio frequency radiation to be detected and interpreted. The dust and gas lanes in the arms of the Milky Way were traced with radio telescopes to provide a better description of the structure of our Galaxy. New objects such as *pulsars* and *quasars* were found. Pulsars are the remains or final stage of stars. Quasars are not fully understood.

Exploration by satellite began in 1957 and opened a new era in astronomy. Only light and radio frequency radiation can pass through the earth's atmosphere to surface observatories. Satellites have detected x-ray sources which are believed to be *black holes* or *collapsars,* the final stage for massive stars. Probes have made encounters and photographed various planets. And, in 1969, astronauts landed on the moon, fulfilling an age-old dream of space exploration.

The Sun

There are very few celestial events available to all observers that can rival the beauty and majesty of the rising sun. In the *Iliad,* Homer paused in the battle between the Greeks and the Trojans to describe the saffron hues of the sky at dawn.

But the sun rising in the east, crossing the sky by day, and setting in the west at dusk reveals little of its true nature. The sun is a star. In fact, it is the only star near enough to be seen well. The sun appears as a sphere of glowing gases, while all other stars are merely points of light. If the sun were located as far away as most bright stars, it would be too faint to be observed without a telescope.

Sunlight is a mixture of all the colors of the rainbow. The blue daytime sky is caused by violet and blue light scattered by air molecules high in the earth's atmosphere. Near the surface, the air contains dust and water vapor which allow orange and red sunlight to pass through more readily than blue, resulting in the familiar red glow of twilight when the sun on the horizon shines through the lower atmosphere.

The motion of the sun across the sky is, of course, illusory. The apparent daily *(diurnal)* westward drift of the sun is a reflection of the earth's *rotation* (axial spin). The earth seems stationary—*terra firma*—with the sky alternating day and night throughout the year.

The Sun's Apparent Diurnal Motion

Observed from above the north pole, the earth rotates *counterclockwise,* from west to east, in a period of 23 hours, 56 minutes, 4 seconds. The observer on the earth shares this motion and sees the objects in the heavens cross the sky from east to west. The exception will be to an observer at the north and south poles, where the celestial objects turn in concentric circles above the horizon.

In the middle latitudes of the northern and southern hemispheres, the sun and stars seem to *rise* and *set.* These terms reflect the old belief that the sky is in motion. What is actually taking place is the lowering of the eastern horizon revealing new objects in the east while the western horizon moves up to cover those in the west.

The rising and setting sun became a convenient timekeeper to regulate activities on the earth. On the eastern horizon at dawn, the sun climbs to its highest altitude by midday and reaches the western horizon at dusk. In the northern hemisphere the *diurnal arc* (daily path) of the sun is to the south; in the southern hemisphere the sun will be north at midday. Between the Tropics of Cancer and Capricorn, the position of the sun at noon will depend upon the time of year and the latitude from which it is observed.

*The rising sun appears flattened
by the refraction or bending of light in the
lower atmosphere near the horizon.*

The Equinoxes and Solstices

The ancients made very careful observations of the sun for religious purposes. This was true for many widely separated cultures such as the Mayan people of Central America, the neolithic builders of Stonehenge in England, and the pyramid designers of Egypt. It was imperative to know the exact movement and position of this sun-god at all times. Temples and other buildings were constructed to determine the exact point of sunrise on the eastern horizon. This might be accomplished with a sight line across two widely separated monoliths or the direction of the shadow cast by an obelisk. These early people discovered that during the year, the sun does not rise at the same point each day. In the northern hemisphere, the sun rises north of east in spring and summer when days are longer than nights and south of east during the short days of autumn and winter. Between these extremes, during planting and six months later at harvest time, the sun rises at the east point on the horizon, and the lengths of day and night are about the same. The changing sun became a god to worship but, more importantly, it provided a means of measuring time over a long period, necessary in an agricultural society.

When the sun is as far north or south as possible, about one week is required to detect any appreciable change in its path across the sky. Here the sun is at the *solstices*. The origin of the word solstice means ''(the) sun (has) stood (still).'' The sun reaches the solstices on June 21 and December 21 in the solar calendar employed in many countries throughout the world. Generally, in the northern hemisphere, the June solstice is called the summer solstice and the December solstice the winter solstice. Since the dates of the seasons in the southern hemisphere are the reverse of those in the northern hemisphere, the terms June and December solstices are considered more appropriate with world-wide application. Since the sun rises north and south of east at different times of the year, at least twice during the year it must be exactly at the east point at dawn. And, in fact, the sun rises in the east twice each year at the time of the *equinoxes*, the first days of spring and of autumn in the middle latitudes. Equinox means ''equal nights'' and denotes a time when the lengths of night and day are the same throughout the world.

Solstices and equinoxes are points on the celestial sphere as well as dates in the year. For example, the March equinox, March 21, refers to a change in season, namely, the first day of spring in the northern hemisphere and the first day of autumn in the southern hemisphere. The March equinox also designates that point in the sky occupied by the sun on March 20 or 21.

At the March equinox, the sun will rise exactly in the east. On that

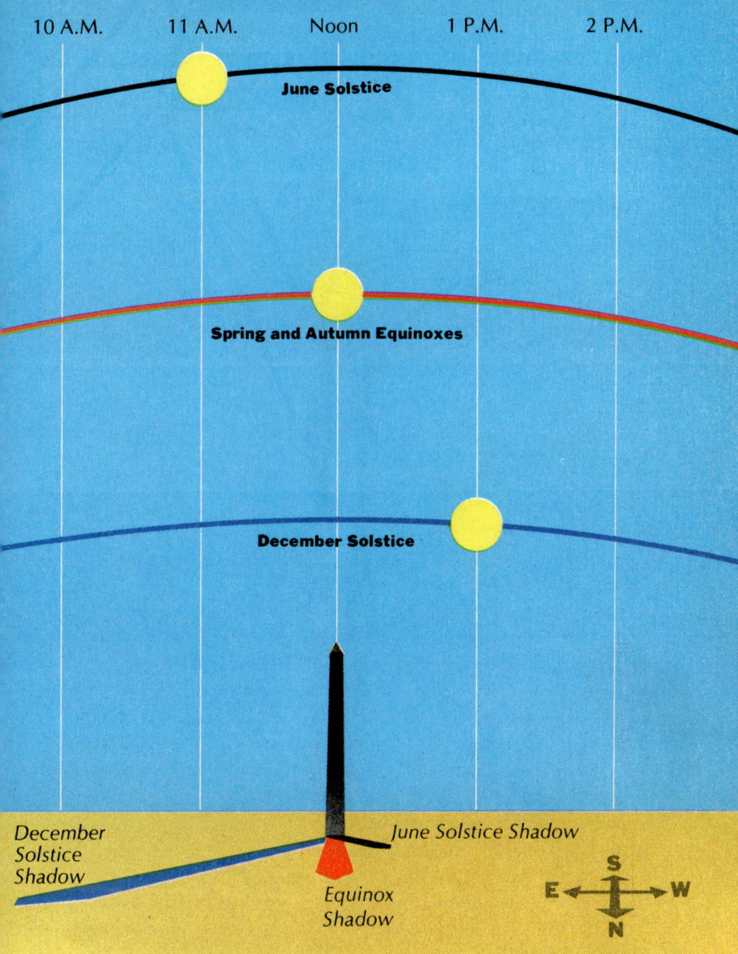

day, the sun reaches the *zenith* (the point exactly overhead at midday) directly over the equator and sets exactly in the west at dusk. In the middle latitudes, the sun will trace its daily path from east to west between the zenith and the horizon. In the northern hemisphere the sun appears to pass south of the zenith, while in the southern hemisphere it will seem to be to the north. Viewed from the north or south pole, the sun will seem to follow the horizon without rising or setting for the entire period of the earth's rotation.

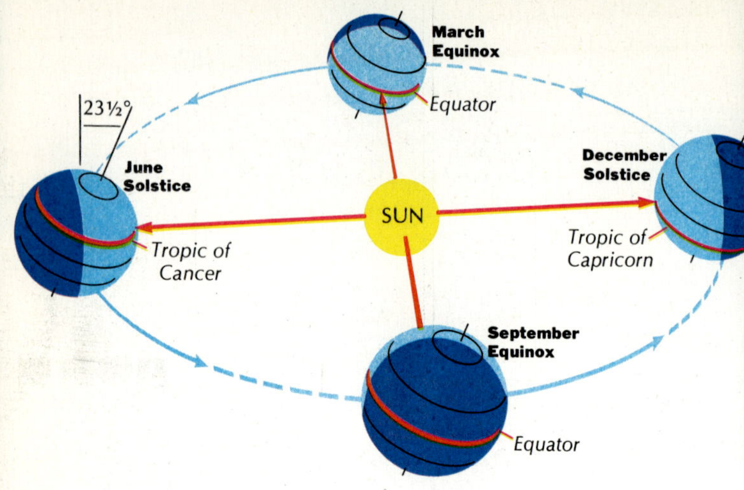

The Seasons

The inclination of the earth's axis to the orbital plane and its revolution around the sun cause the apparent motion of the sun north and south of the equator. On June 21, the north geographic pole is inclined 23½° from the perpendicular to the orbital plane in the direction of the sun. On the day side, the equator is below the plane of the earth's orbit so that the direct rays of the sun strike the earth at the *Tropic of Cancer*, 23½° north of the equator. Summer commences in the northern hemisphere while winter begins in the southern hemisphere. The *terminator* (the line separating day and night) is always perpendicular to the sun's rays. In the northern hemisphere, the inclination of the axis allows more surface area to be exposed to daylight, creating a long day and a short night. The reverse is true in the southern hemisphere. At the equator, the lengths of day and night remain about the same throughout the year. In the northern hemisphere, during summer, the land area north of the *Arctic Circle* receives 24 hours of daylight, causing the phenomenon of the "midnight sun." At the north geographic pole, the sun remains above the horizon from the March to the September equinoxes.

In the southern hemisphere, the area between the *Antarctic Circle* and the south geographic pole is in darkness at the time of the June solstice. Six months later, at the December solstice, the direct rays of the sun will be at the *Tropic of Capricorn* and summer begins in the southern hemisphere. At the March and September equinoxes the earth's axis points neither toward nor away from the sun, and on those days direct solar rays strike the equator and spring and autumn begin with days and nights equal in length in both hemispheres.

The Horizon Coordinate System

On the earth, places and positions are determined by a coordinate system which measures the distance in degrees of arc north or south of the equator and east or west of a great circle (Prime or Greenwich Meridian) at 90° to the equator. Similar grids are used in determining the positions of the sun, moon, and stars. The sky is considered as a huge, infinite sphere above the observer, with these objects on the inside surface of this sphere.

The first celestial grid is called the *horizon system of coordinates* generated about the position of the observer. The observer is at the center of the celestial sphere. His horizontal plane meets the sky at the horizon and cuts the celestial sphere into two parts. The *zenith* is the point directly overhead, at 90° from the horizon. The *nadir* is the corresponding point on the opposite side of the celestial sphere. A line from the zenith to the observer, to the nadir, passes through the center of the earth. *Vertical circles* from the zenith intersect the horizon at 90°. The vertical circle passing through the north point on the horizon, the zenith, and the south point is called the *celestial meridian*. All observers on the same earth meridian share the same celestial meridian, but their zeniths will differ with their distances from the equator. The positions of objects in the horizon system are measured in *azimuth* and *altitude*. Azimuth is the angle along the horizon measured from the north toward the east. Altitude is the angle an object makes with the horizon measured along its vertical circle.

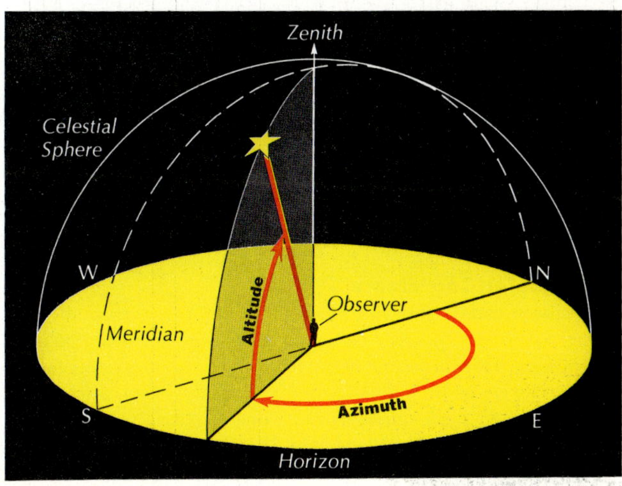

The Celestial Meridian

The *celestial meridian* is defined as that vertical circle on the celestial sphere that passes through the north and south point on the horizon, dividing the sky into eastern and western hemispheres. The meridian can be thought of as the observer's meridian of longitude on the earth extended to the celestial sphere. As the earth turns, the observer and his celestial meridian are carried toward the east. As a result, the stars seem to drift westward to transit, or cross, the celestial meridian. Meridian transits are important in timekeeping and in determining positions of objects on the celestial sphere.

Two additional points are found on the observer's meridian. These are the north and south celestial poles. If the imaginary axis of the earth's rotation were extended into space, it would meet the celestial sphere at two points called the *north* and *south celestial poles*. As the earth rotates on its axis, the sky appears to turn on these poles.

The location of the celestial poles along the meridian depends upon the observer's latitude (angular distance north or south of the equator). At the geographic north and south poles, the corresponding celestial pole is in the zenith. If one extends the earth's equator to the celestial sphere and divides the celestial sphere by a great circle, called the *celestial equator*, midway between the celestial poles, this celestial equator will coincide with the horizon at the poles. At the earth's equator, the celestial equator will extend from the east point through the zenith to the west point, with the celestial poles on the north and south points at the horizon.

In the latitudes closer to the earth's equator, the celestial equator and the poles will be elevated above the horizon an amount determined by the latitude of the position. For example, the *zenith distance* of the celestial equator is the angle along the meridian between the zenith and the celestial equator. This corresponds to the latitude of the position. The angle of the zenith distance is complementary to the angle of the *altitude*. The altitude of the celestial equator is the angle made between the celestial equator and the horizon measured along the meridian. The celestial equator intersects the meridians at right angles.

The Equator Coordinate System

The horizon coordinates measuring altitude and azimuth of the stars pertain to only one position on the earth. This limitation can be overcome by generating a coordinate system fixed to the stars rather than to the observer on earth. In the *equator coordinate system*, the celestial equator becomes the fundamental circle on the celestial sphere midway between the celestial poles. Twenty-four hour circles intersect the celestial equator at right angles and converge at the celestial poles. The celestial equator and the hour circles are fixed to the celestial sphere. As the earth rotates, a star on a given hour circle will transit the meridian once each day.

The angle north or south of the celestial equator is called *declination*. A star north of the equator will have plus (+) declination while another south of the equator will have minus (−) declination. The declination of a star is measured in degrees of arc from the celestial equator along the hour circle passing through the star. *Right ascension* (R.A.) is the measurement in hours, minutes, and seconds of time along the celestial equator toward the east. The origin or zero point is the March Equinox where the ecliptic and the celestial equator intersect. On March 20 or 21, as the sun crosses the celestial equator, its coordinates are right ascension zero hours (R.A. $= 0^h$); declination zero degrees (dec. $= 0°$). At the June solstice, the sun's R.A. $= 6^h$ and dec. $= +23°.5$; at the December solstice, R.A. $= 18^h$, dec. $= -23°.5$. For the present, the stars will be considered to be fixed on the celestial sphere, with permanent measurements in right ascension and declination.

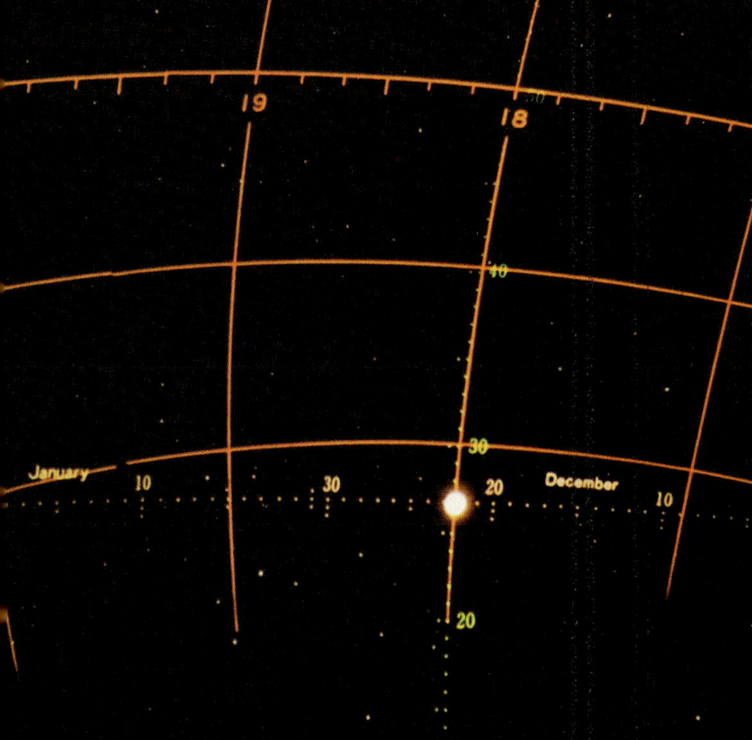

The Sun's Right Ascension and Declination

During the year, the sun traces the ecliptic on the celestial sphere. The earth's axis of rotation is inclined about 23½° to the *ecliptic axis* which is perpendicular to the plane of the orbit. As a result, the ecliptic and the celestial equator form two intersecting circles on the celestial sphere separated by an angle of 23½°. The points of intersection are the equinoxes and the points of greatest separation are the solstices.

As the earth revolves, the sun will appear to move easterly along the ecliptic. With the sun at the March equinox, on the celestial equator, its right ascension (R.A.) will be 0^h and its declination (δ) 0°. The sun continues along the ecliptic toward the east, increasing in right ascension and positive (+) declination. By the June solstice, the R.A. of the sun will be 6^h and its declination +23½°. The sun will cross the celestial equator again at the September equinox. The R.A. is 12^h and δ is 0°. At the December solstice 23½° south of the celestial equator, the sun's R.A. is 18^h and declination negative or −23½°.

In one year the sun will return to the March equinox with R.A. 0^h and $\delta = 0°$. The earth's revolution causes the sun to cross all the hour circles, changing its right ascension through 24 hours. The inclination of the earth's axis is responsible for the change in the sun's declination between + and −23½°.

Solar and Sidereal Time

The earth's rotation provides a means of timekeeping by the sun as well as by the stars. The time required by the sun to make two consecutive meridian transits is called an *apparent solar day*. The time required by a star to make two consecutive meridian transits is a *sidereal day*. *Solar time* is kept with the sun. *Sidereal time* is measured with the stars.

Although these two methods of timekeeping are based upon the rotation of the earth, a solar day and a sidereal day differ in length. Revolution causes the sun to appear about one degree of arc to the east each day. Therefore, the earth must turn an additional degree of arc or four minutes of time to bring the sun to the meridian the following day. A solar day is four minutes longer than the earth's rotational period.

Unlike the sun, the stars are not displaced in right ascension and declination by the earth's revolution. Sidereal time or star time is reckoned by the transit of the March equinox each rotational period. Since the sun crosses the meridian four minutes later each day, the solar day and sidereal day are not compatible. On March 20 or 21, the sun and the equinox transit together. In six months, the sun at the September equinox transits 12 hours after the March equinox. A year will pass before the sun transits together with the March equinox again. Sidereal time measures the rotational period of the earth.

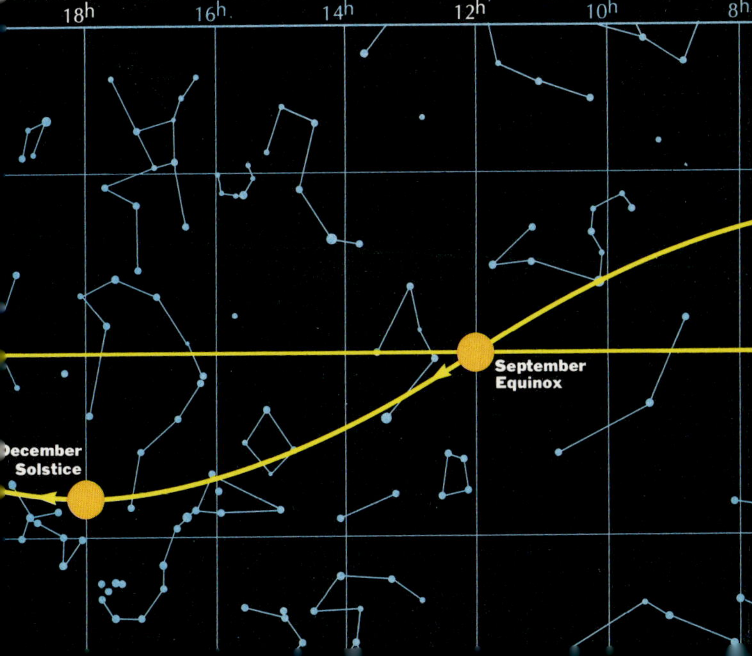

The Stars

To appreciate the stars, the heavens should be viewed on a clear, dark night in the absence of all artificial illumination. At first glance, the sky is a mass of bright, jeweled confusion but a careful investigation reveals patterns and groupings called *asterisms*. Depending on the latitude of the observer, these may include the *Pleiades, Big Dipper* or *Plough*, the *Southern Cross*, the *Great Square*. On a star map, asterisms and nearby stars are connected together to outline imaginary pictures called *constellations*. The *Milky Way* appears as a hazy band of light from millions of stars. The dark rifts in the Milky Way have been found to be extensive clouds of dust and gas. Star clusters or even distant stellar systems are seen as faint patches of light among the constellations. In the southern hemisphere, the *Magellanic Clouds* look as though they are detached portions of the Milky Way, but these are galaxies, the nearest star systems beyond the Milky Way. In the northern hemisphere, a huge spiral of billions of stars called the *Andromeda Galaxy* appears on the threshold of vision in the constellation of Andromeda, between Pisces and Cassiopeia.

Stars are not all alike. Even a casual glance will reveal differences in color and brightness, with brightness being the most obvious distinction. The brightest stars are called first magnitude. An observer with perfect eyesight, viewing the sky on a dark night, will be able to see stars of the sixth magnitude—100 times fainter.

Color differences are more difficult to distinguish than brightness. At first, all the stars look white. Then as they *scintillate* or twinkle in the atmosphere, all the colors of the rainbow seem to radiate from a single star. Like sunlight, starlight is a mixture of all the colors of the spectrum. Each star has a distinctive color determined by its temperature. The sun, like many other stars, is yellow; but there are stars of red, orange, white, green, and blue. A cooler star appears red while a hot star will be blue in color. The sun is an average star between these extremes.

The stars in the direction of the ecliptic (the great circle formed by the plane of the earth's orbit with the celestial sphere) are the constellations of the zodiac. Familiarity with this region is important, for here are found the planets, which may be mistaken for bright stars. During the month, the moon passes through the zodiac as it phases around the earth.

The Milky Way in the direction of the stars in Cygnus. The Veil Nebula is a remnant of a stellar explosion.

A World View

Imagine the sky of the observer in space. Without a horizon, the stars would appear in all directions. The entire celestial sphere could be viewed at one time. In the absence of an atmosphere, the stars would be spectacular—more beautiful than on earth under the best conditions. More than likely, the stars would seem to move, but the observer would realize that this is an *apparent* motion caused by his own spin, much the same as the spinning earth causes night and day.

If he is in the solar system, the sun will also drift into view. The astronaut will revolve in an orbit around the sun in a period of time determined by his distance from the sun. If he is at the earth's distance, he will revolve in one year. As he travels in his orbit, the sun will change its place among the stars. For an earthbound observer, the sun appears to move eastward as the planet revolves in its orbit, and new stars appear in the night sky as the seasons change. An astronaut in space has an advantage over the observer on earth, for there is no horizon in space. The earth itself creates the horizon by cutting the observer's view of the celestial sphere in half, and the spinning earth creates the illusion of a sky turning about two celestial poles. All the stars seem to move in concentric circles about the celestial poles. An observer at the north or south geographic pole will find the center of rotation of the celestial sphere directly above him at the zenith, with all the stars turning parallel to the horizon.

At the earth's equator all the stars rise and set. The celestial poles lie on the north and south points on the horizon. Here all the stars from pole to pole are visible at some time during the year. In the middle latitudes of the northern and southern hemispheres some of the stars are *circumpolar* or always above the horizon, while other stars will rise and set. As the observer travels north or south from the equator (0° latitude) along a meridian, more and more stars will become obscured below the equatorial horizon while more and more will appear above the polar horizon.

The Stars of Orion

Orion, the Mighty Hunter of mythology, consists of seven bright stars that outline the figure. Three of these stars in a row represent his belt. The northernmost star in the belt, *Mintaka*, lies very near to the celestial equator. As the earth turns, this star will trace the celestial equator across the sky. The latitude of the observer will determine how Orion will appear. At the equator, Mintaka rises at the east point perpendicular to the horizon. As the earth turns, the star climbs to the zenith. If Mintaka is rising at sunset, the star will be in the zenith by midnight, setting in the west as the sun rises in the east. In the middle latitudes of the northern

24

hemisphere, all the stars of Orion are visible in the east shortly after the constellation rises. Mintaka is due east. As the earth rotates on its axis Orion travels in an oblique circle above the southern horizon to set in the west. The stars *Betelgeuse* and *Bellatrix* appear higher above the southern horizon than Mintaka. In the middle latitudes of the southern hemisphere, Mintaka rises due east, crosses the sky toward north as it traces the celestial equator. Now the stars *Rigel* and *Saiph* are higher above the northern horizon than Mintaka.

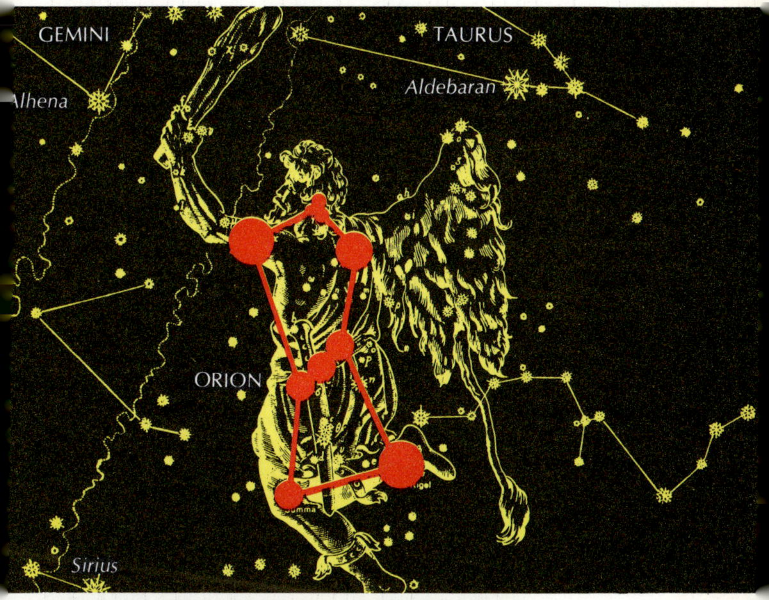

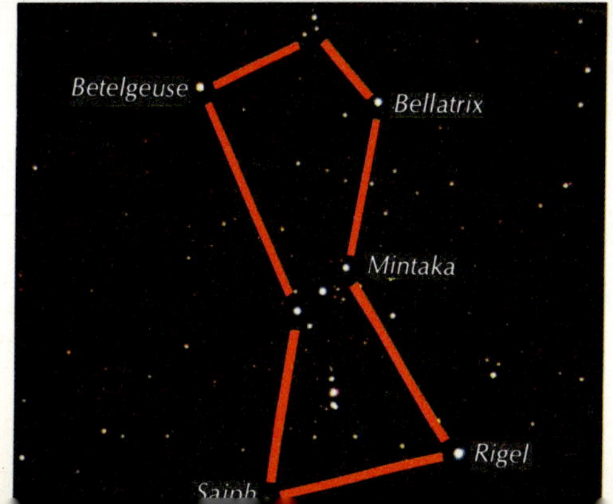

Pisces—Aries

The constellations of the zodiac lie in the direction of the *ecliptic*, the apparent path of the sun. As the earth revolves in its orbit, the sun appears to move easterly about one degree per day. During the course of the year, the sun will pass through these 12 constellations. In addition to the sun, the moon and planets are within the boundaries of the zodiac. Therefore, the zodiac stars were given special recognition by the ancient astrologers. The March equinox is the point in *Pisces*, the *Fishes*, where the ecliptic and celestial equator intersect. Here on March 20 or 21, the sun stands in the zenith on the geographic equator on its way north to the Tropic of Cancer.

Pisces is an extensive constellation portraying two fish tied together by cords attached to their tails. The cords are joined at the fourth-magnitude star, *Alrescha*. The *Western Fish* is formed by a distinctive asterism called the *Circlet*. The *Northern Fish* is represented by four faint stars.

The March equinox is still referred to as the *First Point in Aries*, remindful of the time when the intersection of the ecliptic and the celestial equator was located in the constellation of *Aries*, the *Ram*. *Hamal*, *Sheratan*, and *Mesartim*, the three brightest stars in the constellation, form an obtuse triangle representing the head of the ram, with the rest of the animal outlined by a few faint stars. A great deal of imagination is required to visualize the Ram.

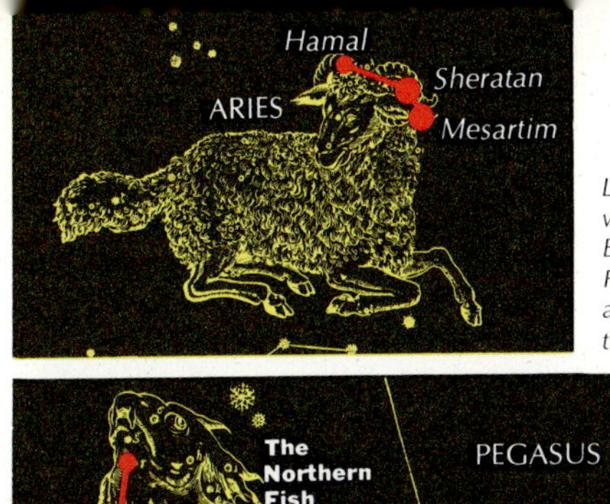

Left: Aries, the ram
with the golden fleece
Below: Pisces, the
Fishes, were astrologically
associated with
the ancient Hebrews.

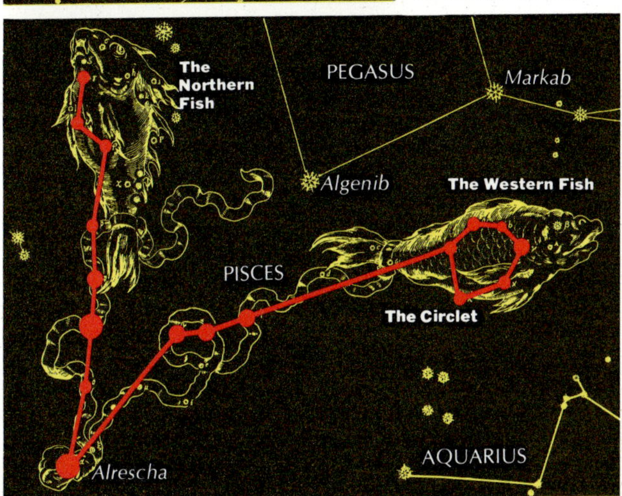

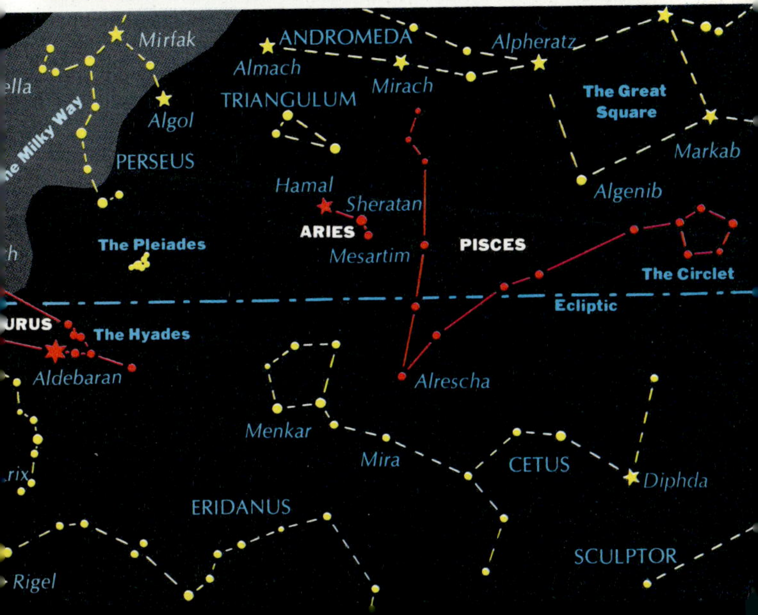

Taurus

Taurus, the *Bull,* contains some of the most interesting stars observed during the entire year. About 4,000 years ago, the March equinox was located between the horns north of the V marking the face of the Bull. The ancients believed that the conjunction of the sun and the sacred Bull was responsible for the fertility of the earth during spring in the northern hemisphere, and the worship of the golden calf was recognition of this astrological belief.

The brightest star in the constellation is *Aldebaran,* an orange star representing the eye of the Bull. This star is found in the V-shaped asterism called the *Hyades.*

The most familiar stars in Taurus are the *Pleiades,* popularly referred to as the *Seven Sisters.* Another cluster, the Pleiades were the mythological daughters of Atlas and the half-sisters of the Hyades. Six stars are seen with ease; the seventh is a test of good seeing. (To astronomers, "good seeing" is a relative measure of the transparency of the atmosphere and not a comment on the visual acuity of the observer.) The second brightest star is called *Elnath* and represents the tip of the horn north of the ecliptic. On star maps, Elnath is shared with the constellation of Auriga, the Charioteer.

Gemini

The stars of *Gemini* outline the figures of *Castor and Pollux,* the *Heavenly Twins.* As the northernmost constellation of the ecliptic, Gemini marks the location of the June solstice, the position of the sun on the first day of summer in the northern hemisphere and of winter in the southern. Castor and Pollux, the two brightest stars, have been considered twin-like since ancient times; yet the stars are not identical: Castor is white, while Pollux appears straw-yellow, and the difference is quite pronounced. The two stars are the heads of the Twins, with the remaining stars grouped in two rows toward Taurus. The constellation terminates in four stars representing the feet.

Gemini is another constellation important in the folklore of the Mediterranean peoples. According to the Greeks, Castor and Pollux shared in the Argonauts' search for the Golden Fleece, represented by Aries, the Ram. Gemini became the patron constellation of seafarers, who relied upon these stars for protection against storms at sea.

The stars of Gemini are conveniently placed to help find the ecliptic and the June solstice. An imaginary line through the star *Wasat* to *Tejat* and *Propus* follows the ecliptic. From Propus, a distance westward equal to the separation between Tejat and Propus marks the northernmost declination reached by the sun on June 21, overhead when observed from the Tropic of Cancer.

28

Top lt.: Taurus, the Bull; stars in clusters, such as the Pleiades and Hyades, have a common origin. Top rt.: Gemini, the Twins; here, in 1781, Herschel discovered the planet Uranus.

RIGA
ON
TAURUS
GEMINI

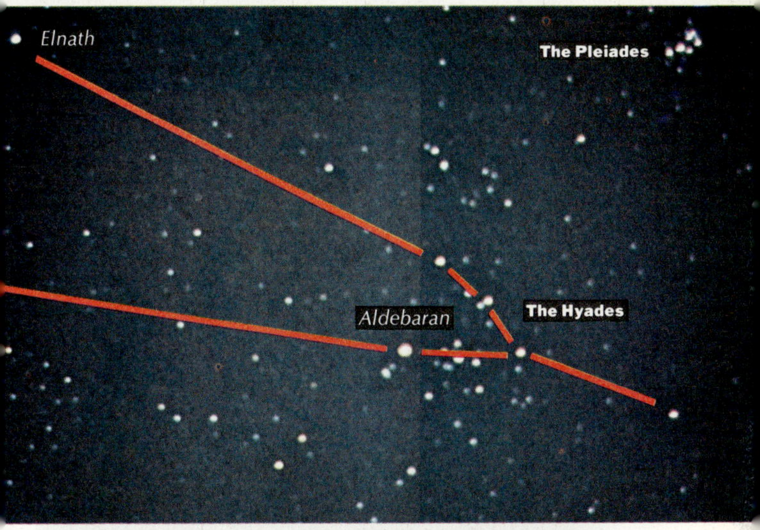

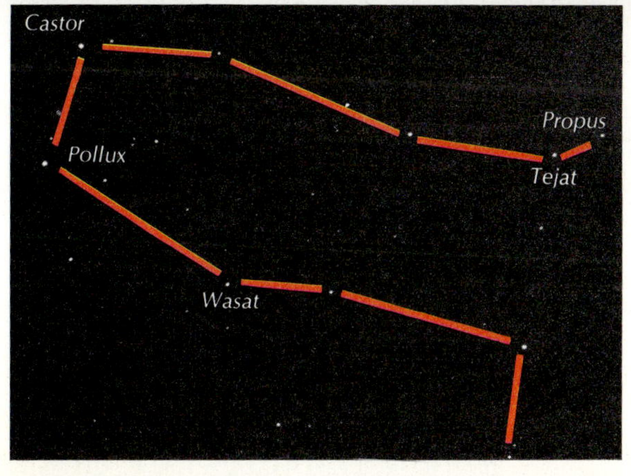
Elnath
The Pleiades
Aldebaran
The Hyades
Castor
Pollux
Propus
Tejat
Wasat

Cancer—Leo

Cancer, the *Crab*, is east of Gemini and the next constellation of the zodiac. Cancer is formed by three stars outlining the body of the Crab and two more representing the claws. Since the stars are faint and difficult to find, first locate the next constellation to the east, *Leo*, the *Lion*. Cancer lies between the bright stars Castor and Pollux in Gemini and *Regulus* in Leo. One of the stars in Cancer, *Asellus Australis,* marks the ecliptic and can be used to trace this imaginary line from Gemini to Leo. The stars *Asellus Borealis* and Asellus Australis refer to the Donkeys present at the birth of Jesus. The star cluster *Praesepe* represents the Manger as well as the Beehive. Binoculars are needed to resolve these stars. It was 2,000 years ago, when the June solstice was located in Cancer, that the Tropic of Cancer was named: it locates the latitude of the overhead sun 23½° north of the geographic equator.

Leo is by far a more interesting group to observe. The head of the Lion is formed by a very distinctive asterism called the *Sickle in Leo*. The hindquarters are in the form of a right triangle. Regulus, identified by its blue color as a very hot star, is the heart of the Lion and is important as a navigational star, lying close to the ecliptic. *Denebola* is the bright star in the tip of the Lion's tail.

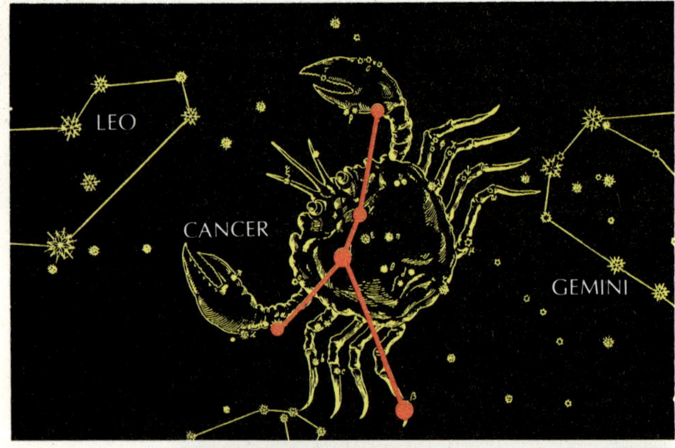

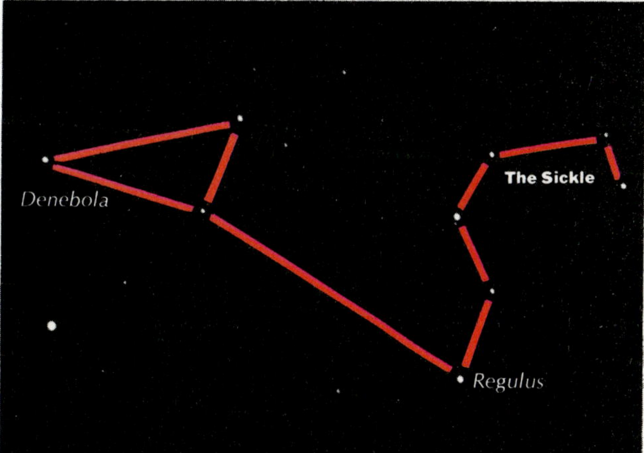

Top: Cancer, the faintest zodiacal
constellation. Btm: Leo; Regulus was one of the
four ''royal stars'' of antiquity.

Virgo—Libra

The sun reaches the September equinox in *Virgo* between the fourth-magnitude stars *Zaniah* and *Zavijava*. The entire constellation is faint because Virgo is located far from the rich star fields of the Milky Way. The bright-blue first-magnitude star *Spica* is the exception and is therefore easy to find among its faint companions. Like Regulus, Spica lies near the ecliptic and helps to locate this imaginary line. The September equinox is almost midway between Regulus and Spica, locating the position of the sun as it crosses the equator toward the south.

Although the stars of Virgo are faint, the *Virgin* was important in mythology. As a fertility goddess, Virgo was in the east at dawn during harvest time. Spica represented a ''spike of wheat.'' Virgo also portrays Justice holding *Libra*, the *Balance* or *Scales*. According to legend, Virgo was the last of the deities to leave the earth for the heavens, holding the scales to judge mankind.

Libra is another inconspicuous constellation that is significant only because it lies in the path of the sun. Only the third-magnitude stars called *Zubenelgenubi* and *Zubeneschamali* are of interest. In Arabic, these names mean ''the southern and northern claws,'' alluding to the next constellation, Scorpius, the Scorpion. Zubenelgenubi lies close to the ecliptic; Zubeneschamali is the only green star bright enough to be seen without optical aid.

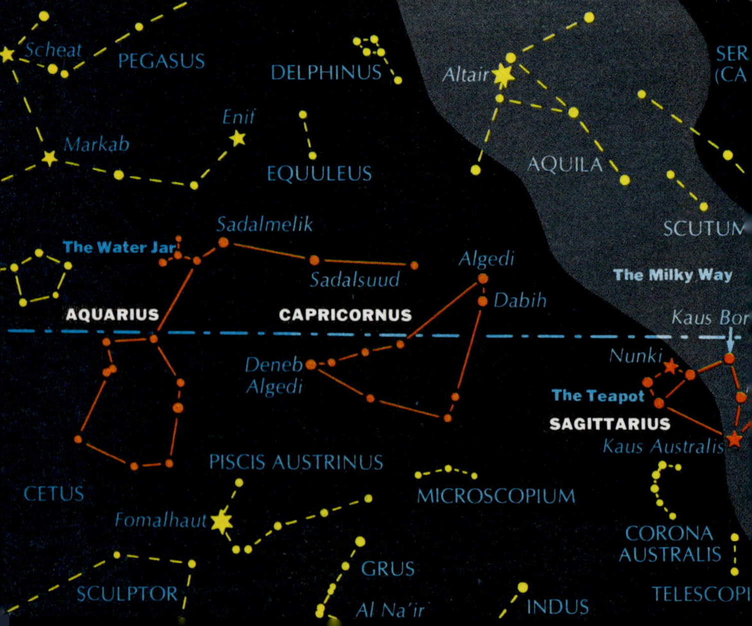

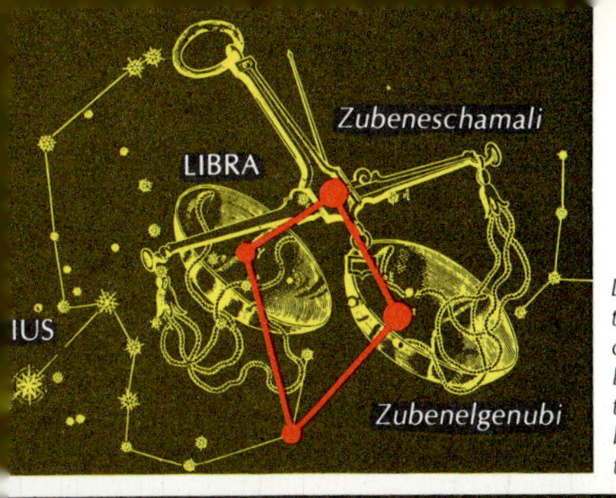

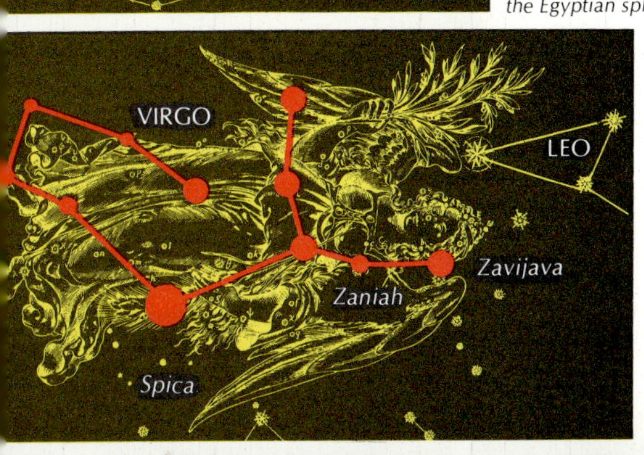

*Left: Libra, the Balance;
the only inanimate
constellation of the zodiac.
Below: Virgo, the Virgin;
the head of Virgo and the
body of Leo form
the Egyptian sphinx.*

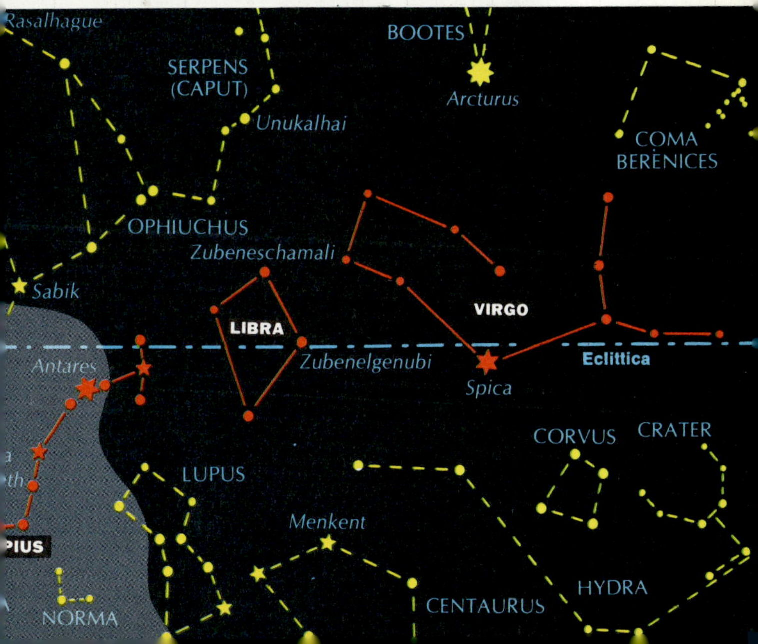

Scorpius

Scorpius, the *Scorpion,* is one of the few constellations that resembles the figure it is supposed to represent. This constellation lies between Libra and Sagittarius, who is aiming his arrow into the Scorpion's red heart, depicted by *Antares,* a first-magnitude red star. Antares means "rival of Mars" (called Ares by the Greeks), and the star is remindful of the red planet. According to mythology, the Scorpion killed Orion, and the gods placed the two opposite each other in the sky to prevent another encounter. Scorpius has many bright stars, most of which are traveling through space in a loose cluster formation called an *association.* The stars of Orion form another example of an association. The claws f the Scorpion extend to Libra but are foreshortened since the original stars of the claws were used to form the Scales. Three stars form the Scorpion's head while the hook-shaped body ends with the second-magnitude star *Shaula* and third-magnitude *Lesath* representing the stinger. The portion of the Scorpion containing the stinger is incorrectly referred to as the "Scorpion's tail," for scorpions do not have tails but an extension of the abdomen.

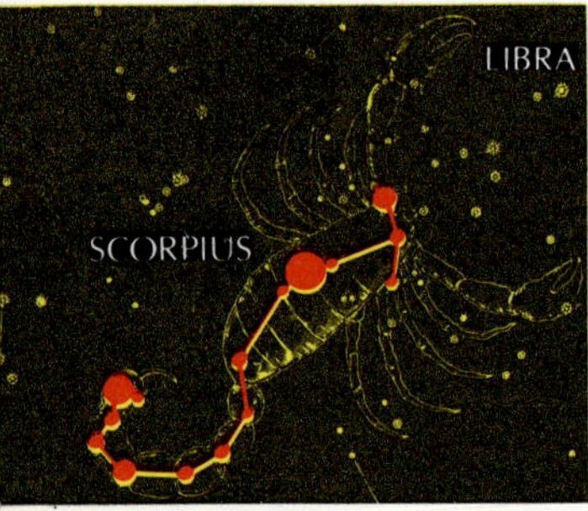

This page: Scorpius, the Scorpion Antares is known as a red giant star. Opposite page: Sagittarius, the Archer, is one of two centaurs in the sky

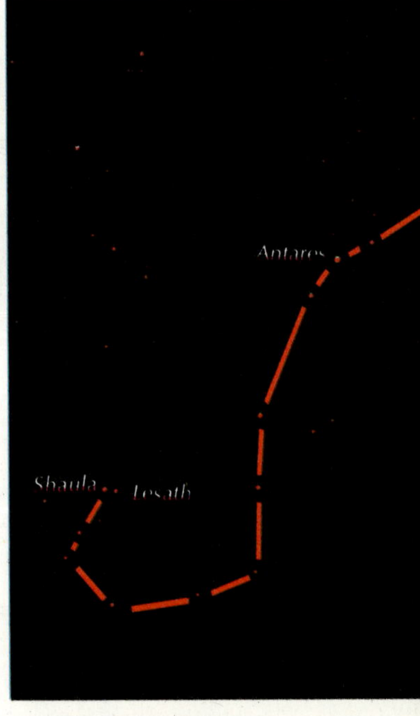

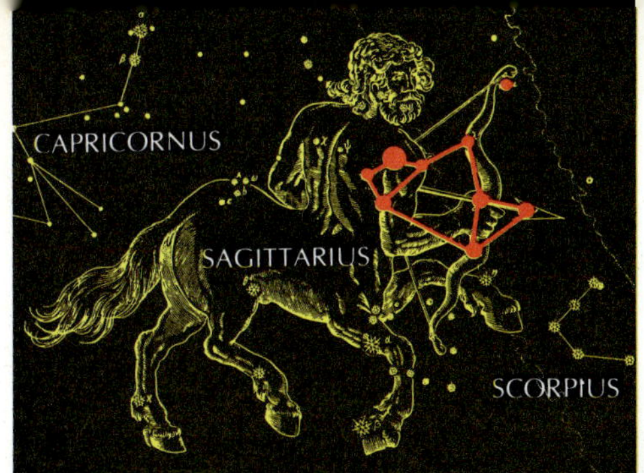

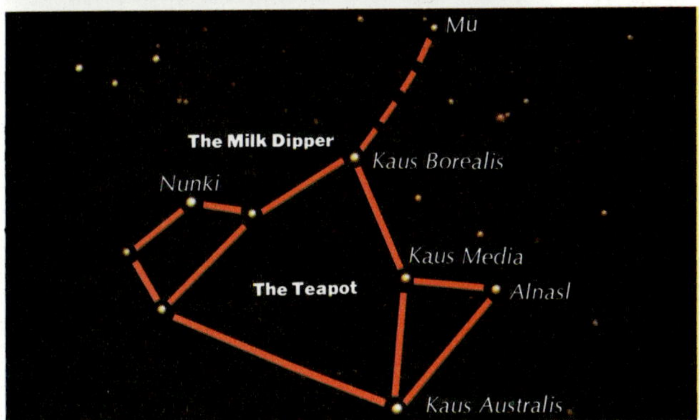

Sagittarius

Sagittarius, the *Archer*, is a centaur firing his arrow into Antares, the heart of Scorpius. Sagittarius is distinguished in several ways. The constellation contains the December solstice which marks the southernmost declination of the sun. When winter begins in the northern hemisphere, summer commences in the middle latitudes south of the equator. The December solstice can be located in the sky by extending a line from the star *Nunki* to *Kaus Borealis* and continuing westward a distance equal to the displacement between these stars.

The constellation forms three interesting asterisms, the bow and arrow of the Archer, the *Teapot*, and the *Milk Dipper*. Kaus Borealis, *Kaus Media,* and *Kaus Australis* make up the bow. Nunki, Kaus Media, and *Alnasl* form the arrow. The Teapot includes all the bright stars in the constellation while the Milk Dipper connects the stars in the handle and lid of the Teapot to the fourth-magnitude star called *Mu* above and east of the December solstice. Sagittarius lies in the direction of the center of the Milky Way and is filled with star clouds and patches of light which are resolved by a telescope into clusters and nebulae.

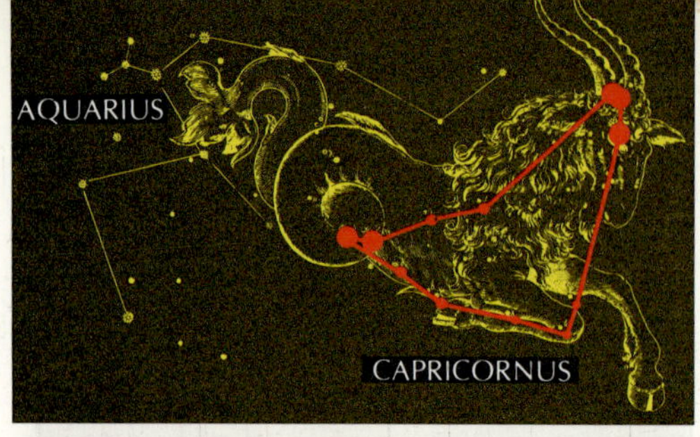

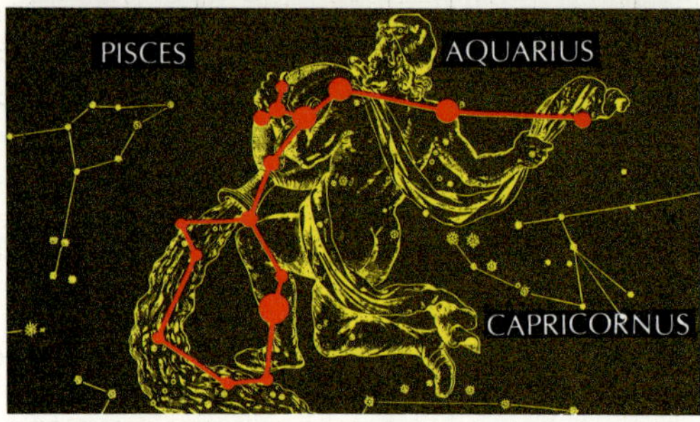

Capricornus—Aquarius

East of Sagittarius, the ecliptic passes through constellations that pertain to water or the sea. These are *Capricornus,* the *Sea Goat, Aquarius,* the *Water Bearer,* and *Pisces,* the *Fishes.* Pisces was described as completing the zodiac, or Circle of Animals. In ancient times, in the Mediterranean regions, the rainy season occurred when the sun was located among these stars, once again reflecting the belief in a direct relationship between celestial motions and events on earth.

Two thousand years ago, the December solstice was located in Capricornus. As a consequence, the geographic latitude where the sun appeared overhead during the solstice was named the *Tropic of Capricorn.* The term is still used although precession has advanced the solstice to Sagittarius. Capricorn, which is depicted as a goat with the tail of a fish, has the form of a large faint triangle considered by the ancients to be the Gate of Heaven. The brightest star, *Algedi,* represents the head of the Goat; *Deneb Algedi* identifies the tail.

Aquarius was considered to be the *Water Bearer* of the gods. The *Water Jar* is a very distinctive asterism formed by a Y of fourth-magnitude stars. Additional faint stars cascade toward the star *Fomalhaut* in the constellation of the *Southern Fish* (Piscis Austrinus). In mythology, these stars represented water pouring from the Jar into the open mouth of the Fish.

     The central star of the Water Jar lies almost on the celestial equator.

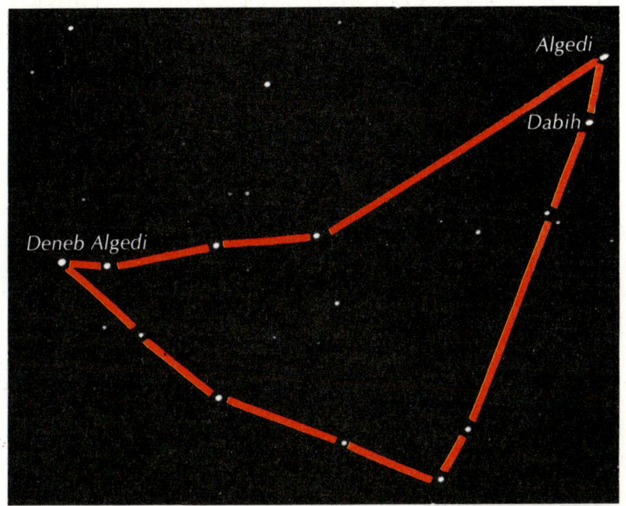

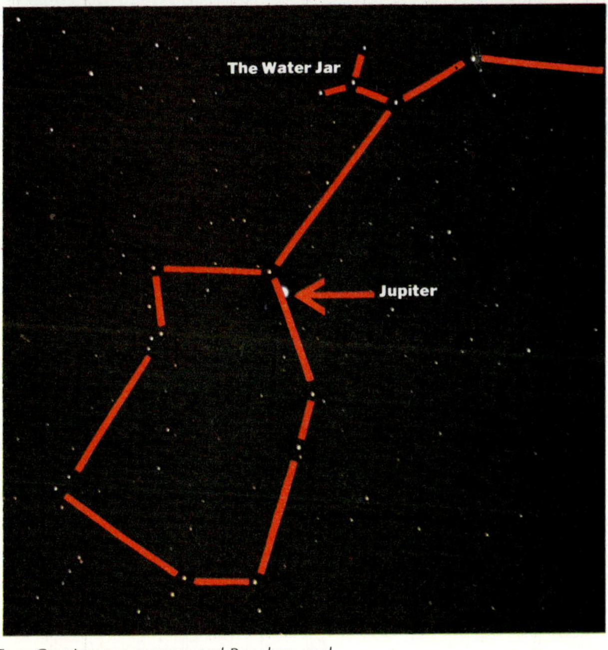

Top: Capricornus represented Bacchus, god
of wine and revelry. Btm.: Aquarius; the planet
Jupiter is the bright object in the center.

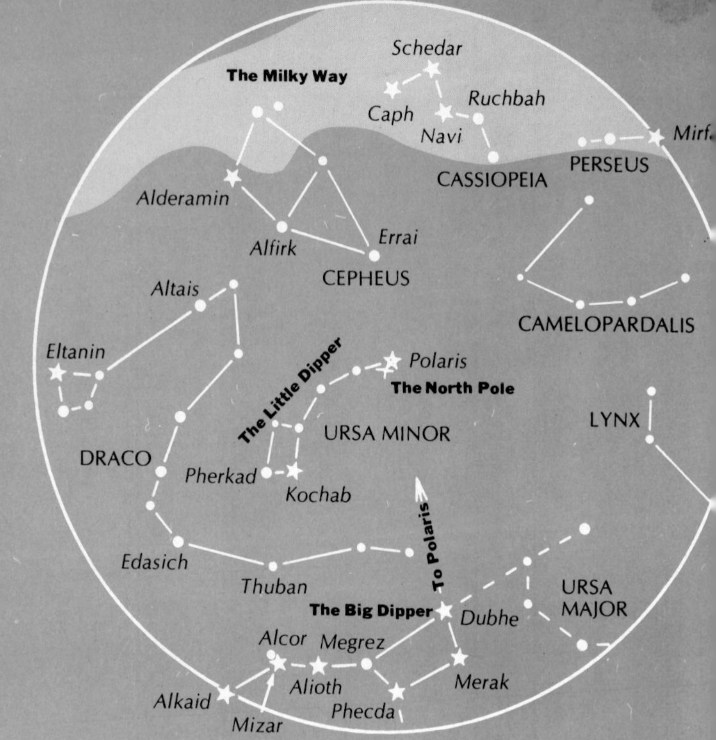

From the Northern Middle Latitudes

To an observer at the north geographic pole, all directions point south. The solstitial colure is in the direction of Betelgeuse, the bright-red star in the constellation of Orion. Betelgeuse will appear to be less than ten degrees above the horizon. *Menkalinan*, the second brightest star in *Auriga*, the *Charioteer*, will lie in the same direction but half the distance between the horizon and the celestial pole. A change in latitude to the south one degree in the direction of Betelgeuse and Menkalinan will bring the stars one degree higher in the sky until, at mid-latitude, halfway between the geographic equator and north pole, Menkalinan will appear in the zenith. All places on the earth with the same latitude will see the sky this way, including France, the United States bordering Canada, southern U.S.S.R., Mongolia, China, and northern Japan. The angle on the meridian between the zenith and the celestial equator is equal to the latitude of the observer. If the celestial equator is 30° south of the zenith, the observer is at 30° north latitude in the southern United States along the border of Mexico, North Africa, the Middle East, and China.

38

This page: Circumpolar stars at 40° north latitude; Opp. page: Star trails show effect of Earth's rotation.

The Northern Circumpolar Stars

In the United States, the most familiar group oт circumpolar stars is called the *Big Dipper*, an asterism known by various names, including *The Plough* and *Charles' Wain*. In Japan it forms the *Emperor's Carriage*. These seven stars form the body and tail of *Ursa Major*, the *Great Bear*. The two stars at the end of the bowl, *Merak* and *Dubhe*, are the *Pointers*, guiding the observer to Polaris and the celestial pole. Polaris marks the tip of the tail of *Ursa Minor*, the *Lesser Bear*. The *Little Dipper* is formed by the seven brightest stars of Ursa Minor. Two of the four stars in the bowl of the Little Dipper are second and third magnitude. They are the so-called *Guardians of the Pole*.

Between the two bears are stars forming the tail of *Draco*, the *Dragon*. The remainder of the constellation curves around the bowl of the Little Dipper toward Polaris. Then the body of the Dragon curves away, back in the direction of the Big Dipper's handle. Halfway back to Alkaid, in the Big Dipper, the Dragon terminates with four stars forming its head; other faint stars represent its fiery tongue.

Opposite the Great Bear on the other side of Polaris, a familiar W or M outlines the chair of *Cassiopeia*, the *Queen*. The star *Caph* lies near the equinoctial colure and can be used as a guide to the March equinox.

Cepheus, the *King*, husband to Cassiopeia, is a faint constellation found between the Queen and the Dragon. One noteworthy feature in Cepheus is the variable star called *Delta Cephei*. It reaches fourth magnitude at maximum and drops down to fifth magnitude at minimum in a period of 5.366 days. Delta Cephei is the prototype for variable stars called *Cepheid variables*..

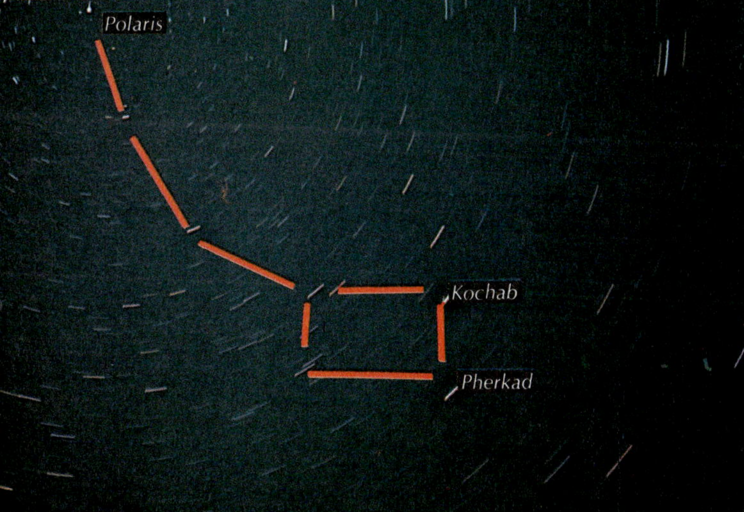

The Northern Sky
in March

Northern sky on March 21
September equinox crosses
meridian at midnight.

The Milky Way
CEPHEUS
Deneb
The
Northern Cross
CYGNUS
DRACO
UR
MIN
LYRA
Vega
The
Little
Dipper
The Big Di
HERCULES
Cluster
CORONA
BOREALIS
BOOTES
EAST
CANES
VENATIC
SERPENS
(CAPUT)
Arcturus
CC
BEREN
Equator
VIRGO
OPHIUCHUS
SCORPIUS
Ecliptic
Spica COR
LIBRA
Menkent
CENTAURUS

Star Magnitudes
-1   0   1   2   3   4   5

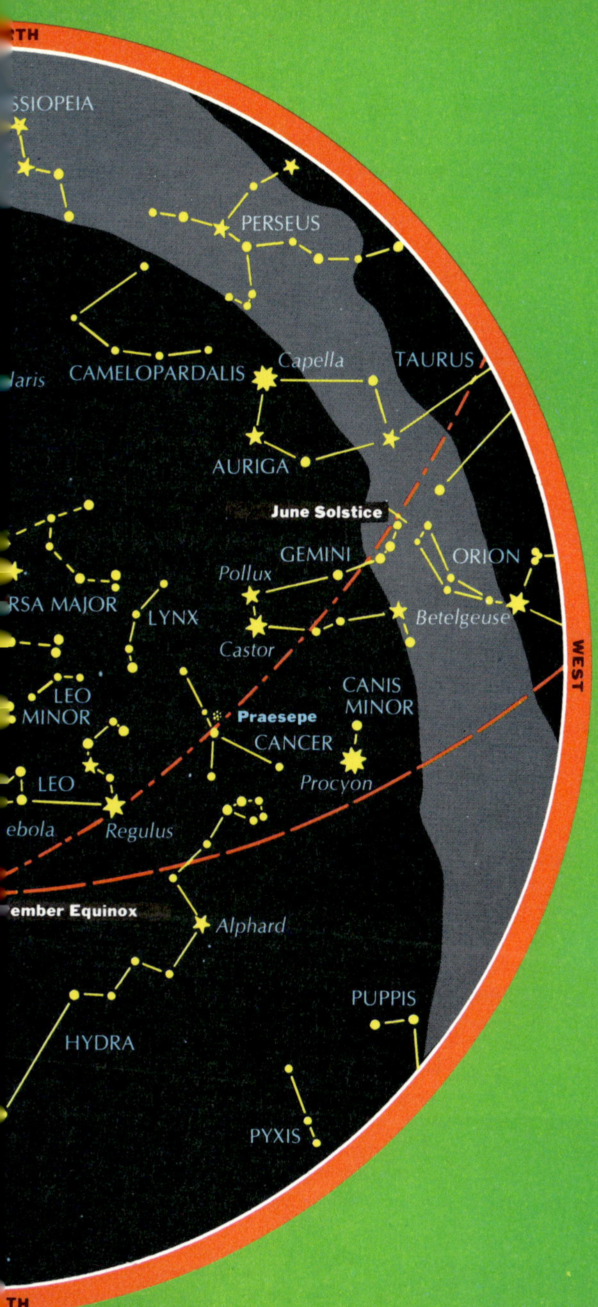
NORTH
CASSIOPEIA
PERSEUS
Polaris
CAMELOPARDALIS
Capella
TAURUS
AURIGA
June Solstice
GEMINI
Pollux
ORION
URSA MAJOR
LYNX
Betelgeuse
Castor
CANIS
MINOR
LEO
MINOR
Praesepe
CANCER
LEO
Procyon
Denebola
Regulus
September Equinox
Alphard
PUPPIS
HYDRA
PYXIS
WEST
SOUTH

The Northern Sky in March

The March sky in the northern hemisphere has very few conspicuous constellations but many bright stars. An exception is the zodiac constellation *Leo*, with the distinctive asterism called the *Sickle*. *Ursa Major* with the *Big Dipper* is high above with *Leo Minor* in the zenith between Leo and Ursa Major. Leo Minor is another constellation construed by Hevelius to fill an area without a named constellation. Here the stars are only fourth magnitude. To the east of Leo and north of Virgo is another faint group called *Coma Berenices, Berenice's Hair*. East of Coma Berenices is a bright-orange star called *Arcturus* in the constellation *Boötes*, the *Herdsman*. Above Boötes and below the handle of the Big Dipper are his hunting dogs, *Canes Venatici*. The Greeks associated Boötes with Arcas, the legendary inventor of the plough, and the constellation was also called *Arktos, Keeper of the Bear*. The star Arcturus retains this ancient name. Arctic Circle, the boundary of the Frigid Zone on the earth, means the circle of the bear. Canes Venatici contains one bright star, *Cor Caroli*, the *Heart of Charles*.

The Northern Sky in June

In June, the stars of Scorpius and Sagittarius sweep low above the southern horizon. Arcturus dominates the western sky while *Vega* in *Lyra*, the *Lyre*, passes overhead. An imaginary line between these stars locates the *Corona Borealis*, the *Northern Crown*, and *Hercules*, hero of the ancient Greeks. The Corona contains a necklace of stars with the brightest called *Gemma*, the *Gem of the Crown*. Gemma is also called *Alphecca*, a corruption of the Arabic word for dish. *Hercules*, or the *Kneeler* forms a large letter H between Corona Borealis and Lyra. In legend, he was noted for his great deeds of strength and was honored by being placed in the heavens. Below the H is the red third-magnitude star, *Rasalgethi*. One of the largest stars, its faintness to our eyes gives little clue to its diameter, which is four times the distance between the earth and sun.

The *Summer Triangle* dominates the sky. This asterism is formed by the brightest stars in three constellations including Vega in Lyra, *Deneb* in *Cygnus*, and *Altair* in *Aquila*. Vega is the brightest star of the summer sky, passing overhead in the middle latitudes and in the zenith between 38° and 39° North Latitude. Altair, in Aquila, the *Eagle* is to the south and is located by its two companions, *Tarazed* and *Alshain* which are equally spaced on opposite sides of Altair. The three stars form a line to Vega. Deneb in Cygnus, the *Swan*, is northeast of Vega. Here another asterism, the *Northern Cross*, stands with its upright member following the Milky Way into the center of the Summer Triangle. The star at the foot of the Cross is Albireo, the Arabic name for "head of the swan."

Rt.: Aquila, the eagle of Zeus
Below rt.: Ursa Major, the Great Bear; Below:
Hercules, hero of the ancient Greeks

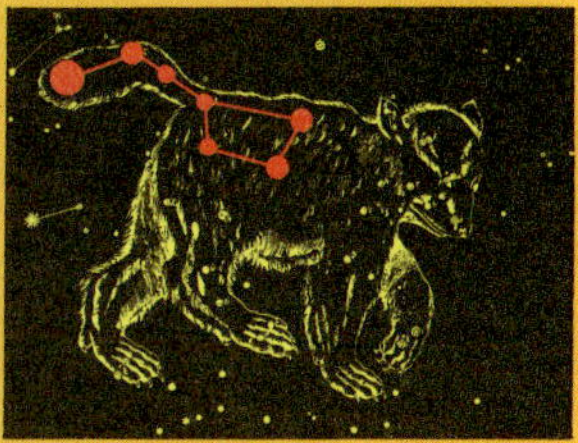

The Northern Sky
in September
Northern sky on
September 22
March equinox crosses
meridian at midnight.
URSA MAJOR
LYNX
Pollux
Castor
GEMINI
Pola
AURIGA
CAMELOPARDAL
June
Solstice
Capella
CASSIOPEI
TAURUS
PERSEUS
Betelgeuse
The Pleiades
ANDROMEDA
EAST
Aldebaran
The
Hyades
Nebula
ARIES
ORION
PISCES
Rigel
March Equir
CETUS
Dipho
ERIDANUS
PHOENIX
Ank
Star Magnitudes
-1   0   1   2   3   4   5

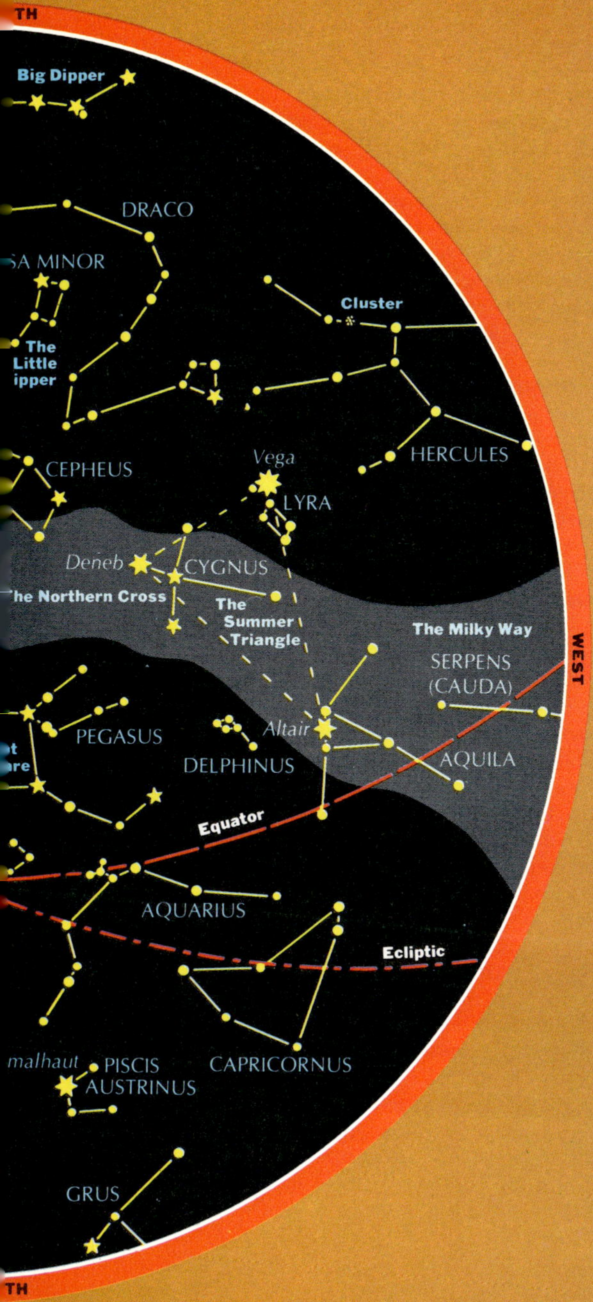

TH
Big Dipper
DRACO
SA MINOR
The
Little
ipper
Cluster
CEPHEUS
Vega
HERCULES
LYRA
Deneb
CYGNUS
The Northern Cross
The
Summer
Triangle
The Milky Way
SERPENS
(CAUDA)
WEST
PEGASUS
Altair
t
re
DELPHINUS
AQUILA
Equator
AQUARIUS
Ecliptic
malhaut   PISCIS
CAPRICORNUS
AUSTRINUS
GRUS
TH

The Northern Sky in September

As autumn commences in the northern hemisphere, the March equinox transits the meridian at midnight. Above *Pisces* is *Pegasus*, the *Flying Horse*, with its distinctive asterism, the *Great Square*. *Alpheratz* in the northeast corner of the square belongs to *Andromeda*, the *Chained Princess*. Two rows of stars forming a wedge to the east outline the figure of the Princess. Above Andromeda is *Cassiopeia*, the *Queen*. *Perseus*, the *Champion*, lies east of Andromeda and extends to *Taurus*. *Triangulum* is south of Andromeda.

The *Summer Triangle* is high overhead at sunset. As the earth rotates, the Milky Way with Cassiopeia and Perseus pass in the zenith. The Great Square stands high in the south. The winter stars of Taurus with the Pleiades are seen on the horizon to the east. The *Great Galaxy* in Andromeda appears as a faint glow above the star *Mirach*. Here is another Milky Way, a stellar system of billions of stars so remote that almost three million years are required for its light to reach the earth.

The legends of the autumn stars are most interesting. Cassiopeia caused her daughter Andromeda to be chained to a rock near a terrible sea monster, *Cetus*. Perseus, who had decapitated the Medusa, releasing Pegasus, the Flying Horse, saved Andromeda by holding the Medusa's head before Cetus. The sea monster turned to stone by gazing at the evil eye, represented by the changing brightness of the star Algol (from Arabic *al ghul*, the Ghoul).

The Northern Sky in December

Orion dominates the December sky of the northern hemisphere. No other constellation has as many bright stars in a distinctive asterism. A halo of bright stars found in other constellations surrounds the "Mighty Hunter." To the ancient Egyptians, Orion was identified with Osiris, who died periodically and was revived by the flooding of the Nile. The three stars marking the belt of Orion, *Mintaka*, *Alnilam*, and *Alnitak*, are useful in locating two more constellations: to the west, the belt stars point to *Taurus*, containing the bright-orange star *Aldebaran*; to the east, the belt stars point to *Sirius*, the *Dog Star*, in *Canis Major*, the *Great Dog*. Sirius is the brightest star in the sky with −1.42 magnitude. A line diagonally through Orion from the blue star *Rigel* to *Betelgeuse* and to the north locates *Gemini* with *Castor and Pollux*. Above Orion, the sky is dominated by *Auriga*, the *Charioteer*. The prominent star in the upper western corner of this pentagon-shaped figure is *Capella*, a star of 0.08 magnitude. North of Auriga lies *Lynx*, a faint constellation with only one star brighter than fourth magnitude. Named by Hevelius in the 17th century, it is too faint to have been included among the classical constellations of antiquity.

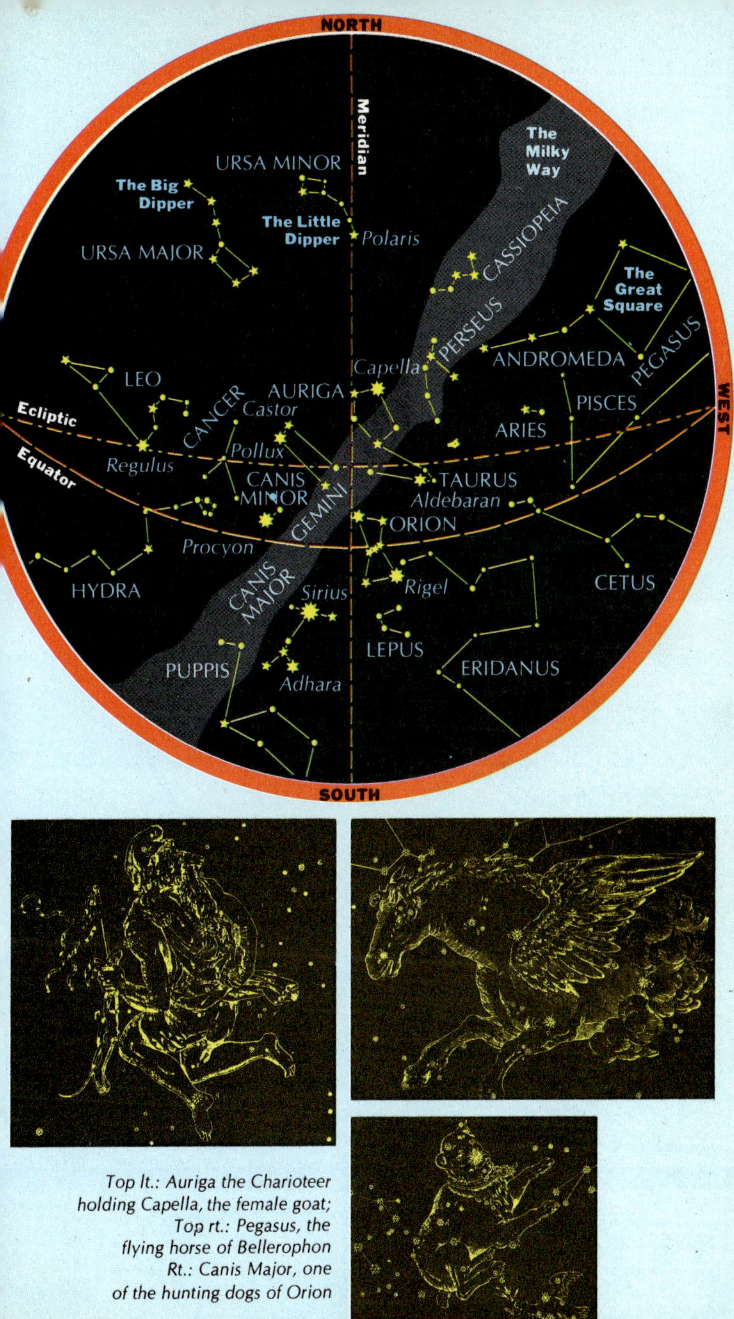

Top lt.: Auriga the Charioteer holding Capella, the female goat; Top rt.: Pegasus, the flying horse of Bellerophon Rt.: Canis Major, one of the hunting dogs of Orion

From the Southern Middle Latitudes

At the south geographic pole all directions point north with the belt of Orion on the horizon. This constellation locates the solstitial colure at 6^h R.A. Facing Orion, *Canopus,* the bright star in *Carina,* the *Keel,* appears halfway to the zenith. Sirius, the brightest star, is about seventeen degrees above the horizon. An imaginary line from Canopus to *Furud* and *Mirzam* in *Canis Major* runs parallel and near to the colure. Moving to the north brings Orion above the horizon. The south celestial pole is depressed southward the same number of degrees of arc as the change in latitude. At 30° south latitude in South Africa, Argentina, and Australia, Orion crosses 60° above the northern horizon. The celestial equator strikes the horizon at the east and west points. The angular distance on the celestial meridian between the zenith and the celestial equator measures the latitude of the position. This angle is equal to the elevation of the celestial pole above the southern horizon. With 6^h R.A. on the meridian, the ecliptic will intersect the celestial equator at the east and west points on the horizon. The June solstice is 36½° above the north point on the horizon.

48

The Southern Circumpolar Stars

At the south geographic pole, all the stars are circumpolar and remain above the horizon during the 24-hour day. At the equator, all the stars rise and set in the same period of time. In the middle latitudes (New Zealand at 40° south), most of the bright stars are circumpolar above the southern horizon. When the solstitial colure is on the meridian, the bright star Canopus approaches the south point on the horizon. A visitor from the northern hemisphere will find that the rotation of the earth causes the sky to appear to turn clockwise, the opposite of the northern stars. Halfway between the south horizon and the celestial pole is the ecliptic south pole, the point on the celestial sphere which is perpendicular to the earth's orbital plane. The angular separation of 23½° between this point and the celestial pole is a measure of the earth's inclination on its axis. Nearby is the Large Magellanic Cloud which has the appearance of a detached portion of the Milky Way. To the northwest lies the Small Magellanic Cloud which, like its larger companion, is another star system beyond the Milky Way.

Crux, the *Southern Cross*, serves the southern skies in the same manner as the *Big Dipper* does the northern. In the Big Dipper, the bowl stars *Merak* and *Dubhe* point to the north celestial pole. The upright member of the Cross, formed by *Acrux* at the foot and *Gacrux* at the head, points in the direction of the south celestial pole. Unfortunately, there is no "pole star" in the southern sky. To find the pole, one must imagine a line in the sky between *Achernar* in *Eridanus* to *Hadar* in *Centaurus*. Extending the upright member of Crux will intersect this line at the celestial pole. As the earth rotates, the Cross circles the pole in 24 hours, with Acrux and Gacrux pointing in its direction.

Southern Milky Way in the direction of Crux; the dark nebula is called the Coal Sack.

The Southern Sky in March
Southern sky on March 21
September equinox crosses meridian at midnight.
Achernar
The S
Magel
C
HYDRUS
RETICULUM
DORADO
The La
Magel
Cloud
COLUMBA
Canopus
CHAMAELEON
CARINA
Adhara
VELA
The
False
Cross
C
CANIS MAJOR
LEPUS
The
Sout
Cros
Sirius
PUPPIS
PYXIS
The Milky Way
HYDRA
Equator
Alphard
Septen
Equ
Procyon
CANIS MINOR
Regulus
GEMINI
Ecliptic
LEO
Praesepe
Pollux
CANCER
LEO MINOR
Castor
LYNX
URSA MAJOR
WEST
Star Magnitudes
-1   0   1   2   3   4   5

ANA
Peacock
PAVO
CORONA
AUSTRALIS
CTANS
ANGULUM
TRALE
Atria
ARA
SAGITTARIUS
A
dar
Rigil Kentaurus
SCORPIUS
LUPUS
December
Solstice
SERPENS
(CAUDA)
Antares
Cluster
Menkent
NTAURUS
OPHIUCHUS
EAST
LIBRA
Spica
US
SERPENS
(CAPUT)
VIRGO
HERCULES
Arcturus
MA BERENICES
CORONA
BOREALIS
NES VENATICI
BOOTES
ig Dipper

The Southern Sky in March

In March at midnight, Regulus in the Lion is above the horizon to the northwest. To the west of Regulus, the head of *Hydra*, the *Sea Serpent*, follows Procyon to the west point on the horizon as its long, twisting body arches above Leo toward the zenith and Libra in the east. *Alphard*, a second-magnitude star, locates the heart of Hydra. The Greeks saw the constellation as the Sea Serpent, while the Egyptians pictured the stars as the Nile River. The triangular head of the serpent became the river delta. The sun was on the ecliptic north of Hydra when the Nile flooded its banks. Two small constellations, *Corvus*, the *Crow* or *Raven*, and *Crater*, the *Cup*, are perched on Hydra's back.

The False Cross and Crux are between the zenith and the south celestial pole. *Rigil Kentaurus* and *Hadar* and *Menkent* locate the Centaur standing over the Cross. To the Greeks, Centaurus was Chiron, the wise tutor of Achilles. An unusual feature is *Omega Centauri*, a globular cluster of stars bright enough to be seen with the unaided eye. Globular clusters surround the center of, and provide a clue to, the Galaxy's structure and the earth's location within this system of stars. Toward the pole are faint stars in the constellations *Chamaeleon*, the *Lizard*; *Musca*, the *Fly*; and *Circinus*, the *Compasses*.

The Southern Sky in June

In June, the Milky Way arches high overhead with Scorpius and Sagittarius near the zenith. Canopus is about to cross the southpoint while in the north, the fifth brightest star, Vega, passes above the horizon. Between Vega and Scorpius stands *Ophiuchus*, the *Serpent Bearer*, grasping a huge snake represented by *Serpens Caput*, the *Head*, and *Serpens Cauda*, the *Tail*. This constellation has a legend which associates these stars with Laocoön, the Trojan priest who with his sons was killed by the serpent for warning against the wooden horse brought by the Greeks. The *Corona Australis*, the *Southern Crown*, is in the zenith. This semicircle of stars is remindful of its northern counterpart, the *Corona Borealis*. To the south is a triangular group of faint stars called *Telescopium* which is another of the instruments added to the list of constellations in modern times. Another navigational instrument, *Norma*, the *Level*, is found nearby. Between the Telescope and the Level is *Ara*, the *Altar*. Between the celestial pole and these faint constellations is a region of bright stars. Looking south, Alpha Centauri appears to the west of the meridian. East of the meridian and higher in the sky is *Pavo*, the *Peacock*. Its brightest star is also called *Peacock*. Between these stars and south of Ara, three stars form an isoceles triangle called *Triangulum Australe*, the *Southern Triangle*, with the navigational star *Atria* forming

52 the apex of the triangle.

Rt.: Corona Australis, the Southern Crown
low rt.: Serpens, associated with Laocoön story
elow: Centaurus with Crux, the Southern Cross

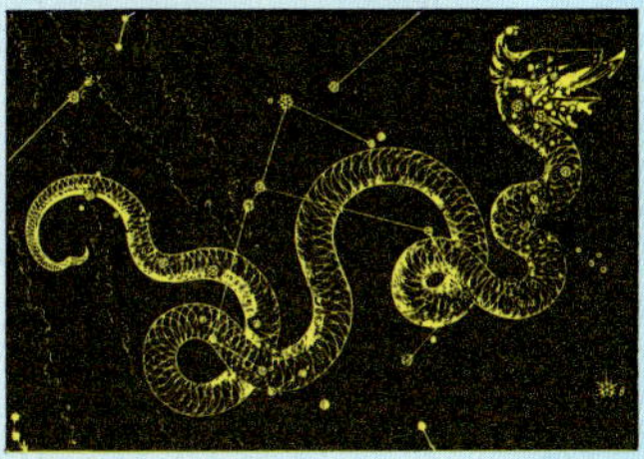

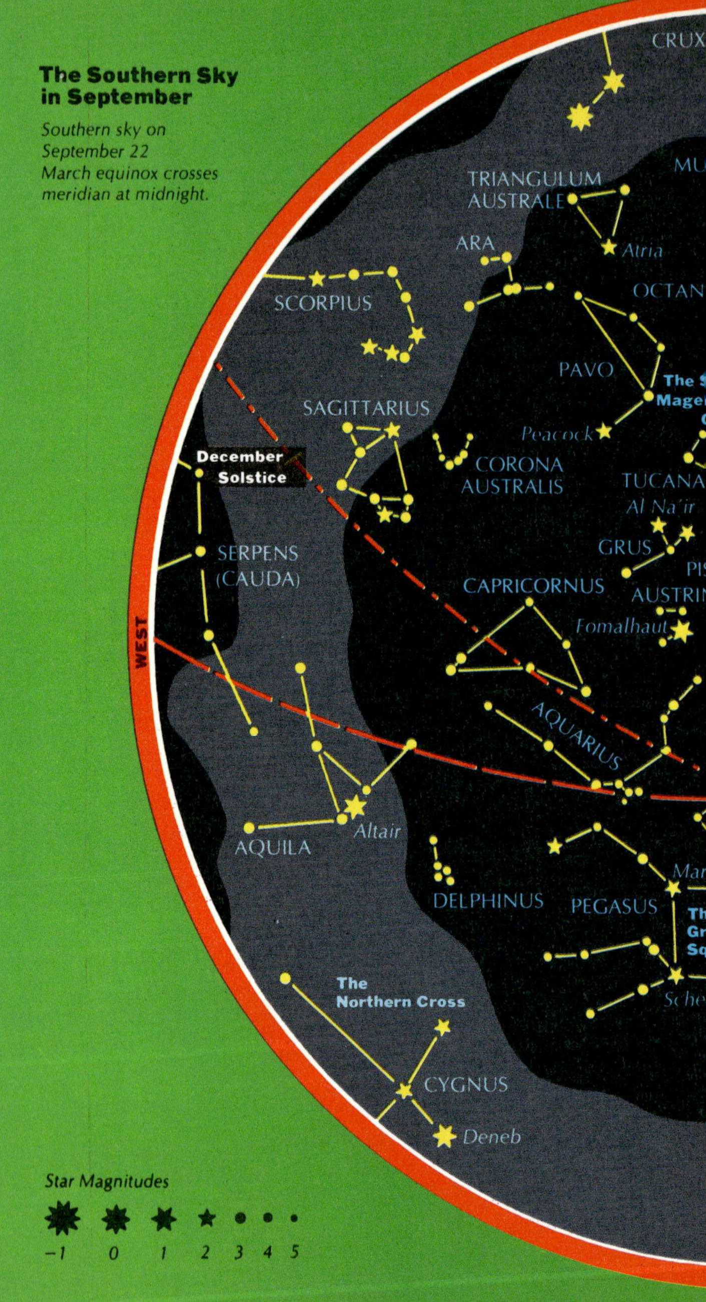

The Southern Sky in September

Southern sky on September 22 March equinox crosses meridian at midnight.

CRUX
MU
TRIANGULUM AUSTRALE
Atria
ARA
OCTAN
SCORPIUS
PAVO
The S Mage C
SAGITTARIUS
Peacock
December Solstice
CORONA AUSTRALIS
TUCANA
Al Na'ir
GRUS
PIS
SERPENS (CAUDA)
CAPRICORNUS
AUSTRIN
Fomalhaut
WEST
AQUARIUS
Altair
AQUILA
Mar
DELPHINUS
PEGASUS
Th Gra Sq
Schea
The Northern Cross
CYGNUS
Deneb

Star Magnitudes
-1   0   1   2   3   4   5

The Southern Sky in September

The September sky contains a large number of constellations associated with water. To the north in the zodiac *Capricornus, Aquarius,* and *Pisces* are identified. Between these constellations and the zenith are *Piscis Austrinus,* the *Southern Fish,* and *Cetus,* the *Whale.* From Rigel in Orion on the eastern horizon, *Eridanus,* the *River,* meanders across the southeastern sky.

Pegasus, the *Flying Horse,* skims the northern horizon at midnight with the Southern Cross at the opposite horizon to the south. A line through *Alpheratz* and *Algenib* in the Great Square continued to Crux follows the equinoctial colure along the meridian. Just east of the meridian and south of Pisces is Cetus with the second-magnitude star *Diphda,* the tail of the Whale. West of the meridian and near the zenith is the first-magnitude star *Fomalhaut,* located by a line from *Scheat* and *Markab* in the Great Square. Fomalhaut and Diphda form a lozenge with *Ankaa* in *Phoenix* and *Al Na'ir* in *Grus.* Grus, the *Crane,* is interesting for the double star, Delta, that forms a triangle with Al Na'ir and Beta. Eridanus winds and turns over a large part of the sky and suggested a river to several Mediterranean cultures. The Small Magellanic Cloud lies on the edge of a large triangle of third-magnitude stars representing *Hydrus,* the *Water Snake.*

The Southern Sky in December

The southern December sky contains the brightest stars seen from the earth. The brightest of all is *Sirius,* in the constellation of *Canis Major.* This star is high in the north at midnight and forms a line with the three stars in the belt of Orion. According to the Greeks, Canis Major was one of the hunting dogs of Orion. In fact, the constellation was visualized as a dog by several cultures. In Egypt, Sirius was observed to rise with the sun at dawn shortly before the flooding of the Nile River, and this *heliacal rising* was believed to be responsible for the inundation.

Canis Minor, Orion's other hunting dog, is identified by *Procyon,* the *Little Dog Star.* Procyon is so named because it rises before Sirius. Just above the horizon are *Castor* and *Pollux,* the Gemini. High in the zenith is *Canopus,* in the constellation *Carina.* Canopus is the second brightest star and may appear slightly fainter than Sirius but is intrinsically brighter with a much higher temperature. *Carina,* the *Keel,* was a part of a larger constellation called *Argo Navis,* the ship of the Argonauts of Greek legend. Two stars in Carina, *Epsilon* and *Iota* and another pair, *Kappa* and *Delta* in *Vela,* the *Sail,* form the "False Cross" which has been mistaken for Crux much to the dismay of mariners. Other parts of the old constellation of Argo include *Pyxis,* the *Compass,* and *Puppis,* the *Stern,* of the ship.

Above: Hydra, the Sea Serpent, legendary multi-headed creature
Left: Cetus, the Whale or Sea Monster. The variable star Mira at the base of Cetus' neck is invisible most of the year.

The Moon

The moon is the second brightest object in the sky. In times before the development of artificial illumination, the full moon provided the light to brighten the hours of darkness. The full moon nearest to the autumnal equinox, the *harvest moon,* aided farmers by extending the time they could work the fields after sunset. In winter, the short days are offset by the bright, full moon high overhead. At a mean distance of 239,000 miles, the moon is the nearest object in space. The earth and moon together revolve in orbits around the sun. Generally, the moon is considered to be a satellite of the earth, but since the earth is only four times larger in diameter, the two are sometimes called a "twin planet" system. In its revolution, the moon changes its position toward the east about 13° per day, and, as it revolves, the amount of its surface visible from the earth changes. This *phasing* of the moon contributes to its beauty and fascination. The word *month* is derived from a "moonth" of time or the days required for the moon to pass through all its phases.

Years ago the moon was believed to have influence upon people. The word *lunacy* meant "possessed by the moon." Werewolves were unfortunate people who were transformed into beasts at full moon.

Details of the moon's surface cannot be observed without optical aid. The light and dark markings that outline the "man in the moon," "lady in the moon," or the "hare," are regions differing in smoothness. The bright areas are covered with craters while the dark are flat lava plains, once believed to be large bodies of water, hence called *maria,* the plural of *mare,* the Latin word for "sea."

Above: According to a 19th century hoax, the moon was inhabited by batmen. Opp.: The moon was worshipped as a goddess of light of the night sky.

The Waxing Moon

At new phase the moon is in conjunction with the sun. The nightside of the moon is in the direction of the earth. Most of the time the moon passes north or south of the sun at new phase. In those rare instances when the sun and moon are in direct line with the earth, the moon blocks the view of the sun, resulting in a solar eclipse.

The moon advances eastward and in a day or two appears over the western horizon as a thin crescent in the twilight glow. The age of the moon is reckoned from the new phase. Two days after new phase, the moon is said to be a two-day-old waxing (increasing) crescent. At this time the light from the bright dayside of the earth falls on the dark nightside of the moon, and the entire face of the moon shines with a soft ashen glow called "earthshine." The moon reaches first quarter phase in about 7½ days after new phase, with half of the bright face toward the earth. In the northern hemisphere, the moon will be above the south horizon at first quarter phase. In the southern hemisphere, the moon will be seen to the north. The ten-day-old moon is gibbous or convex on both sides, with most of the face bright with sunlight. Fifteen days after new phase, the moon is opposite the sun and rises in the east at sunset. Now the entire dayside of the moon is visible. As the earth turns, the full moon rises, crosses the sky, and reaches the western horizon at dawn. The rising and setting moon appears orange as its reflected light passes through the earth's twilight glow. The full moon also appears copper colored during a lunar eclipse when sunlight passes through the earth's lower atmosphere before reaching the moon.

The Waning Moon

At age fifteen days, the moon has completed half of its journey back to new phase. From full to last quarter phase, the moon rises between sunset and midnight. The waning (decreasing) crescent reaches the eastern horizon in the early hours before dawn. The moon wanes as it continues in its orbit approaching the direction of the sun.

A few days after full phase, the waning gibbous moon stands above the horizon in the southwest at sunrise. On the moon, the sunset terminator—the line separating the bright dayside from the night— slowly sweeps across the familiar side facing the earth, for, like the earth, the moon is rotating on its axis to alternate day and night. The location of the moon in the sky determines how much of the lunar dayside is visible from earth. Revolving eastward, the moon reaches last quarter at age 22 days. With the rising sun in the east, the last quarter moon will be south for observers in the northern hemisphere and over the northern horizon in the middle latitudes of the southern hemisphere.

As the days pass, the moon continues its sur.ward journey. Now the terminator curves toward the bright limb to form the late, or waning crescent. Once again, earthshine brightens the darkened, sunless areas of the moon. Since the same side is always facing the earth, the nearside never is as dark as the farside, which always faces away from our planet. At age 29½ days, the moon passes between the earth and the sun at new moon phase. When conjunction occurs, the age of the moon is zero days, and the moon is lost in the glare of the sun with the side of the moon toward the earth completely in the shadow.

Opp.: The waxing moon from early crescent to full phase; Above: The waning moon from full to late crescent phase.

The Apparent Orbit

From our planet, the moon appears to revolve in an elliptical curve with the earth stationary in space at one focal point of this ellipse. As we know, the earth revolves in its own orbit around the sun; therefore, the so-called orbit of the moon is not a closed curve since the moon cannot return to the same point in space around the moving earth. It is a *mean apparent* orbit that changes shape from month to month. For example, the nearest approach to the earth by the moon, called *perigee*, is 221,463 miles; the *apogee*, or greatest distance from the earth, is 252,710 miles at maximum. During any one month, these minimum and maximum distances are not necessarily reached because perigee and apogee vary with each revolution.

The lunar orbit is inclined to the ecliptic by an angle of about five degrees of arc. The two points of intersection between the orbit of the moon and the plane of the ecliptic are the nodes. The line of nodes connecting these points drifts westward along the ecliptic for one revolution in a period of 18.6 years. The major axis of the elliptical orbit is called the *line of apsides.* This line connects perigee and apogee. The line of apsides makes one complete turn to the east in about nine years. Seen from a point above the orbit, the moon seems to revolve counterclockwise from west to east. The speed in orbit varies with the moon's distance from the earth. Therefore, the eastward motion of the moon will be greater at perigee than apogee, averaging to about 13° per day.

The time required by the moon to complete one revolution in its orbit and to pass through all its phases is not the same. If the earth were stationary in space, one lunation and revolution about the earth would be similar in length. Since the earth revolves too, once around the earth, a *sidereal month,* is shorter in length than a lunation, a *synodic month.*

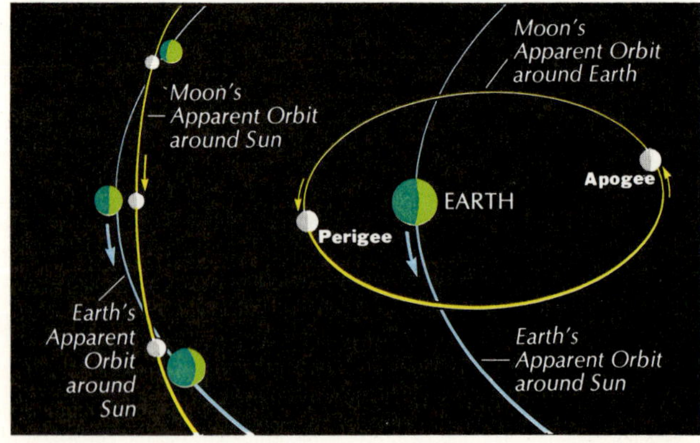

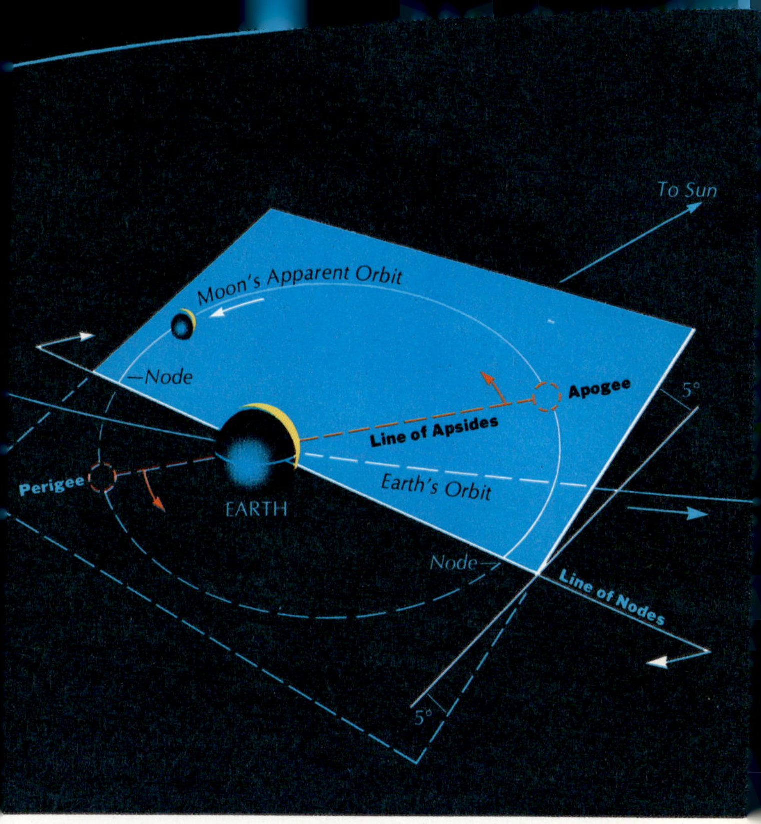

*Above: The moon
appears larger at perigee
than at apogee.*

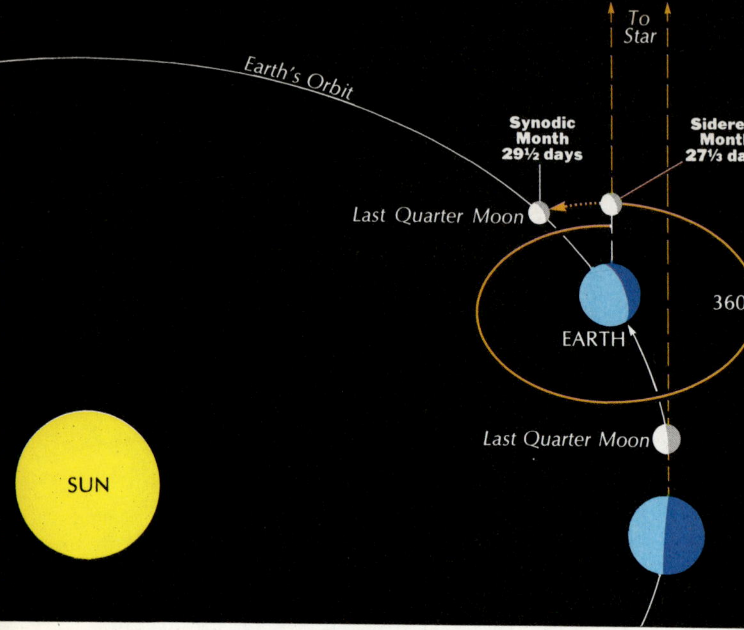

Sidereal and Synodic Periods

A *sidereal month* is the interval of time required for the moon to complete one revolution and return again to the same position among the stars. Consider the moon and a star crossing the celestial meridian at the same time. The following night, the moon will have advanced 13° to the east and will transit about 50 minutes later than the star. In 27⅓ days the moon and the star will again be in line. Although the moon has completed 360° and has returned to its starting point, an additional two days of revolution will be required for it to return to the same phase. The month of the phases or the *synodic month* is 29½ days in length. Picture the last quarter moon on the meridian in conjunction with a bright star at dawn. Each morning the moon will appear closer toward the direction of the sun. By new phase the moon and sun will be in conjunction.

As the earth revolves in orbit, the sun moves eastward about one degree per day (see page 13), so that stars will appear displaced to the west by the same amount. Each day at dawn, the bright star that was in conjunction with the moon will be one degree farther west. At the end of one sidereal period, the moon and star will be in conjunction 27° west of the meridian. To complete the synodic period, the moon must return to the meridian and last quarter phase. Since the moon revolves about 13° per day, an additional two days' orbital motion will complete the

64 synodic period of 29½ days.

Rotation

The moon rotates on its axis in a period of 27⅓ days, which is equal to its revolution in one sidereal month. This is called *synchronous rotation,* and it keeps the same side of the moon facing in the direction of the earth. The familiar face of the "man in the moon" is always turned toward the observer on earth but never the back of his head. If the moon did not rotate, all of its surface would be visible from earth during one sidereal month. The effect of synchronous rotation would be interesting to observe from the moon rather than the earth. Viewed from the moon, the sun and stars appear to slowly drift westward, while the earth remains almost stationary in the sky. The motion of the stars from east to west is caused by the rotation of the moon on its axis. The combined effects of rotation and revolution keep the earth hovering in the same position while passing through phases—like those of the moon observed from earth—but in reverse order. The earth shows a slight displacement east and west as the moon accelerates between apogee and perigee.

Picture the sun and stars on the meridian in conjunction with the earth. The nightside of the "new" earth is toward the moon. After one sidereal period or one rotation, the stars will return again in the direction of the earth on the meridian. But the sun will be 27° east of the meridian, the earth, and the stars. This displacement of the sun to the east is caused by the revolution of the moon and earth around the sun. Two more days will pass before the sun joins the earth on the meridian. The sun and earth are in conjunction in a synodic period. The earth and stars reach two consecutive conjunctions in a sidereal period.

*Above: Crescent earth
as seen from Apollo 17 in
orbit around the moon.*

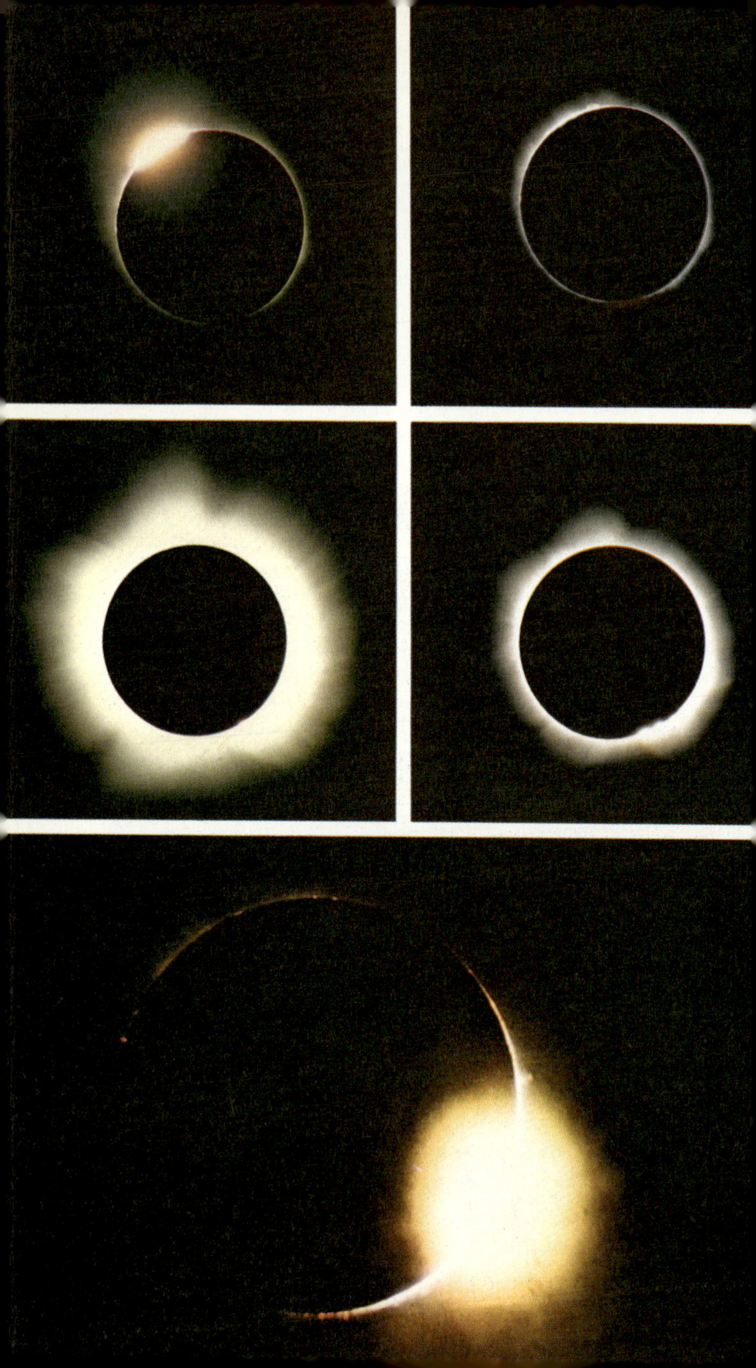

Solar Eclipse

Perhaps the most spectacular of all celestial events is an eclipse of the sun. On these occasions the new moon passes in front of the sun and covers the bright disk. Suddenly the sky becomes dark enough for bright stars to be seen. The temperature drops ten or more degrees, and a strange, eerie specter appears in the sky. In a few moments, the sun returns and all is normal again. It was a frightening experience for early man who thought a demon or dragon had consumed the sun. He may have accidentally looked up to see the partial phase of the eclipse already underway. When the sun disappeared, a black hole was seen surrounded by glowing, bright, nebulous streamers.

The dark disk is the nightside of the moon and the pearly glow is the *corona* which is the outer atmosphere of the sun. The corona can be seen only when the bright disk of the sun is occulted, either artificially with instruments like the *coronagraph* attached to a telescope, or naturally by the moon during an eclipse. Eclipses of the sun are rare and cannot occur more than four times each year. Not all eclipses are *total,* with all of the sun hidden from view; some are *partial,* when the new moon and the sun are not exactly in line with a point on the earth, and though the moon reaches conjunction, it cannot cover the entire face of the sun.

Another type is an *annular eclipse* where the moon appears too small to obscure the entire disk of the sun and the sun is seen as a bright ring or annulus. During an annular eclipse, the moon is at or near apogee and has a smaller angular displacement in the sky, and the earth must be near or at perihelion—its closest approach to the sun; thus, the sun will have a larger angular displacement than the moon. Since

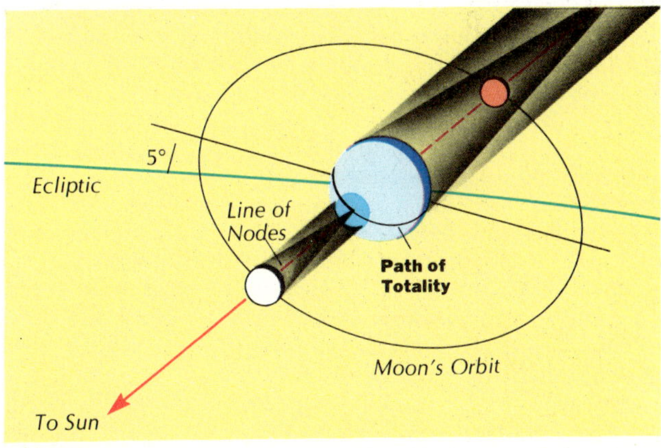

Opp.: Total eclipse of the sun, showing the corona or outer atmosphere; the "diamond ring" is formed an instant before and after totality.

perihelion occurs in January, annular eclipses are more frequent at that time. Total eclipses occur in July when the earth is at aphelion. An eclipse cannot occur every new moon phase because the orbit of the moon does not coincide with the ecliptic plane. At an angle of about 5°, the moon crosses the ecliptic twice each month at two points called nodes. When the sun is at or near a node at new moon, an eclipse of the sun can occur. At other times, the new moon either passes above or below the sun.

The moon casts a conical shadow in space opposite the sun. By coincidence, the length of the shadow is about as long as the distance between the earth and moon. At total eclipse, the apex of the shadow strikes the earth as a small dark disk only a few miles across. As the earth turns and the moon revolves in its orbit, the shadow traces a narrow *path of totality* across the surface of the earth. Totality can only be observed in this narrow band.

Lunar Eclipse

Everyone on the nightside of the earth can see a *lunar eclipse* all at the same time. Unfortunately, lunar eclipses occur less frequently than solar eclipses, the maximum number in any year being three. Like the moon, the earth has a long shadow opposite the sun. The shadow extends a distance of over 800,000 miles. The moon is about 250,000 miles from the earth and is much smaller in diameter than the earth's shadow at that distance, hence the moon can be eclipsed if it is at or near a node at full moon phase.

Since a line from the centers of the sun and earth lies in the ecliptic plane, the centerline of the earth's shadow is on the ecliptic plane. When the moon is full and opposite the sun it can enter the shadow of the earth. To do so, one of the moon's nodes must be in the shadow. Once in the shadow, the moon will lose the direct light from the sun. But the moon does not become dark and disappear altogether. Light from the sun strikes the earth and passes into the atmosphere. The long wavelengths of light that produce red and orange colors scatter in the air, painting the sky with the familiar hues of sunrise and sunset. Refraction bends this light into the shadow where it falls on the face of the moon. As the moon enters the shadow, it takes on a mysterious coppery color and remains that way until it passes out of the shadow.

The ancients believed that dire events were forecast on these occasions. In England, Stonehenge may have been a luni-solar observatory for the purpose of determining the positions of the sun and moon during the course of the year. The positions of these rocks may have served as a ''computer'' to predict in advance the awesome phenomena of solar and lunar eclipses.

68

Top: The full moon eclipsed in the earth's shadow. Btm.: Predictions of solar and lunar eclipse may have been made at Stonehenge as well as observations of the sun at the June solstice.

The Planets

Planets and stars can be identified in several ways. Usually, a planet appears as a brighter object shining with a steady light among the stars of the zodiac. The reflected sunlight from planets reaches the earth in a bundle of rays or a beam and is therefore less affected by atmospheric motion than starlight which passes through in a single ray. Planets twinkle or *scintillate* less than the stars. The most reliable method of identifying planets is to become familiar with the star patterns in the zodiac constellations. The orbits of the planets follow the ecliptic and are therefore seen against the background of these stars. A bright addition to an asterism or a constellation will immediately be identified as a planet.

Mercury is nearest to the sun and can only be observed at dawn or dusk as a morning or evening star of −1.9 magnitude, somewhat brighter than Sirius, the brightest star. Venus, at −4.4 magnitude the brightest planet, is also between the earth and sun. In the sky, Venus can be as much as 47° from the sun and may set as late as three hours after sunset.

Mars, Jupiter, and Saturn revolve in orbits beyond the earth and can be seen all night long. Mars is orange-red in color and is as bright as −2.8 magnitude at its closest approach to the earth.

Usually the planet Jupiter, with −2.5 magnitude, is the brightest in the sky. Saturn is −0.4 magnitude and yellow in color. Neptune and Pluto are too faint to be seen without optical aid. Uranus with +5.6 magnitude is bright enough to be visible to the naked eye.

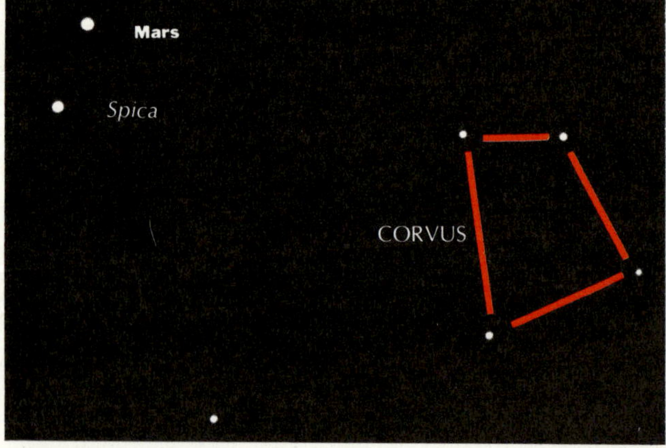

Direct and Retrograde Motion

Planets revolve counterclockwise in *direct motion* from west to east. At *opposition,* when the earth passes between a superior planet and the sun, the planet will be displaced on the celestial sphere from east to west, in *retrograde motion.* After opposition, the planet resumes its easterly course among the stars.

These changes in direction are apparent and due to the differential orbital velocity of the planet and the earth. The effect is most noticeable with Mars, which revolves in an orbit nearer to the earth than that of the other superior planets.

Mars requires 687 days, its *sidereal period,* to revolve around the sun. The time between two successive oppositions is about 780 days, the *synodic period.* Between oppositions, Mars is in *conjunction* on the other side of the solar system, with the sun between the planet and the earth.

At opposition Mars is nearest the earth and reaches its greatest brilliance. Unfortunately, this planet has an eccentric orbit, and opposition distance varies between 36 million and 63 million miles. A favorable opposition occurs when Mars is at or near *perihelion,* its closest approach to the sun (and therefore to the earth). Then the planet appears as a bright-orange, star-like object of −2.8 magnitude.

71

Opp.: The planet Mars near the star Spica in Virgo. Above: The retrograde path of Mars in Taurus and Gemini.

The Milky Way

On a clear, dark night the stars seem too numerous to count. Yet only about 6,000 of the known billions of stars can be seen with the unaided eye. At first, the bright stars attract the viewer's attention. Then a careful study reveals faint stars that twinkle in and out of sight. Orion, for example, overwhelms the stargazer with the seven bright stars that form the asterism of the Mighty Hunter. But many stars within the figure can be counted as they sparkle on the threshold of vision.

The *Milky Way* locates the greatest concentration of these faint points of light as well as many bright stars. In fact, a belt of bright gems follows and almost coincides with the Milky Way as it circles the celestial sphere. The Milky Way lies in the plane of the *Galaxy* which is our star system in space. At first glance it seems as though the sun and the earth are located in the center of a huge flat aggregation of stars. For many years it was believed that the sun did have a preferential place in space, and it was not until the present century that observation confirmed the center of the Galaxy to be in the direction of Sagittarius where the Milky Way appears to be most extensive. Although the Milky Way shows the direction of most of the stars, it does not have uniform brightness along its entire length.

A journey along the Milky Way might begin at the June solstice in the constellation Gemini which lies near the *galactic equator*, the intersection of the plane of the Galaxy on the celestial sphere. Following the galactic equator toward the southeast, the Milky Way passes between Procyon in Canis Minor and Betelgeuse, the bright-red star in Orion, to the faint stars of Monoceros, the *Unicorn*. Continuing southeast of Sirius, the Dog Star, the plane of the Galaxy passes the constellations Puppis, Pyxis, and Carina of the ancient Argo Navis. Meanwhile, the Milky Way is becoming brighter approaching Crux, the Southern Cross. This region is broken by a dark band or rift culminating in the dark cloud near Crux called *The Coal Sack*. At first the great rifts were believed to be sparse regions with a view beyond the stars into distant empty space. Now it is known that the rifts are extensive clouds of dust and gas.

Continuing along the galactic equator, the Milky Way brightens, reaching its magnificence toward Scorpius and Sagittarius. Here the structure seems almost chaotic and no longer follows a narrow band. The diffuse glow broadens into an irregular ball-like structure dotted with puffs of brightness. Proceeding north, the Milky Way passes through Aquila to Cygnus. Here are found dark rifts remindful of the Coal Sack in the southern hemisphere. After Cygnus, the Milky Way enters the constellations of Cassiopeia, Perseus, Auriga and back again to the June solstice in Gemini.

*The Milky Way contains many
bright and dark nebulae, such as the
North America Nebula in Cygnus.*

Other Stellar Systems

The tour of the Milky Way has provided an indication of the vastness of the stellar system to which the sun belongs. There are at least 100 billion stars in this expansive disk which is so large that a light beam would require 100,000 years to cross it. When its size was determined at the turn of the century, astronomers believed that the Milky Way represented the entire universe of stars in an endless expanse of space. But there were other points of view. In the 18th century, Immanuel Kant had proposed that the faint patches of light such as the Andromeda Galaxy might be other star systems beyond the Milky Way. He referred to these as *island universes* in a sea of emptiness.

One of these extragalactic nebulae is visible to the unaided eye. That is the Andromeda Galaxy, which lies toward the plane of the Milky Way. The stars in the constellation of Andromeda are members of our

Galaxy and are merely in the direction of the more distant star system. The Andromeda Galaxy is the only large spiral galaxy visible without a telescope. It is by far the most distant object seen with the naked eye—almost 3 million light years from the earth. In other words, the light now received on earth left the star system almost 3 million years ago. If the Milky Way were a kilometer or about ½ mile across, the Andromeda Galaxy would be another wheel of stars twice that diameter at a distance of 25 kilometers or 15 miles. Much closer to the Milky Way are the irregularly shaped Magellanic Clouds. These are satellite galaxies held by the gravitational attraction of the billions of stars of our Galaxy.

But the eye alone cannot perceive the nature of distant objects. Most of the universe goes unseen. From the earth, only the sun and moon and an occasional comet reveal their physical appearance. The telescope must be employed to extend our vision to more distant wonders in space.

Part 2•Through the Telescope

The 200-inch Hale Telescope at Mount Palomar (below) is the largest reflecting telescope in the United States. The world's largest reflector (236-inch) is in the Soviet Union.

Telescopes

The Refracting Telescope

Optical telescopes are of two basic designs, *refractors* and *reflectors*. Both types concentrate the light at one point called the focus. Here an image is formed and magnified with lenses called *oculars* or *eyepieces*. Refracting and reflecting telescopes differ in the way in which light is brought to the *focal point*.

The first telescopes designed in the 17th century were of the refracting type. Basically, this telescope consists of a large lens called an *objective lens* through which light enters the *telescope tube*. The purpose of the objective lens is to gather light at the focal point. *Light-gathering power* increases as the area of the objective increases. Therefore, the objective lens should be as large as possible. The largest refractor is the 40-inch telescope at the Yerkes Observatory. Compared with a small amateur telescope, the 40-inch gathers 400 times more light than a 2-inch or 100 times more light than a 4-inch refractor.

As light enters the telescope, the curved surface of the lens causes the light to strike the objective at an angle. On entering the lens, the light is *refracted,* or bent, as in a prism. The curvature of the lens bends the light near the edges more than the light entering the center, thereby bringing the rays to a focus. However, light is made up of the seven colors of the spectrum, which bend in varying amounts, so that a simple objective lens cannot bring the light to one focus. The problem is solved by an *achromatic objective* made of two lenses of crown and flint glass. Used together, these lenses diminish *chromatic aberration* and bring the colors closer to one focal point.

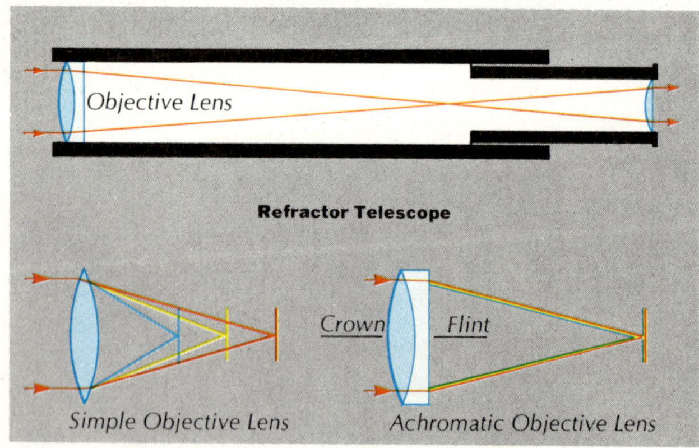

The Reflecting Telescope

A *reflecting telescope* uses a *primary mirror* in place of an objective lens. Light is gathered on a curved mirror at the bottom of the telescope. The mirror is made of a glass disk that has been ground and polished to the form of a paraboloid. The glass is then coated with a fine film of aluminum to produce a reflective surface. The paraboloid has the property of bringing the parallel light rays to a focus. Since the light was not refracted, achromatic aberration is eliminated. Another advantage is size: a reflecting telescope with a shorter focal length can be built larger **79**

Above: The 40-inch refracting telescope at Yerkes Observatory is the largest in the world.

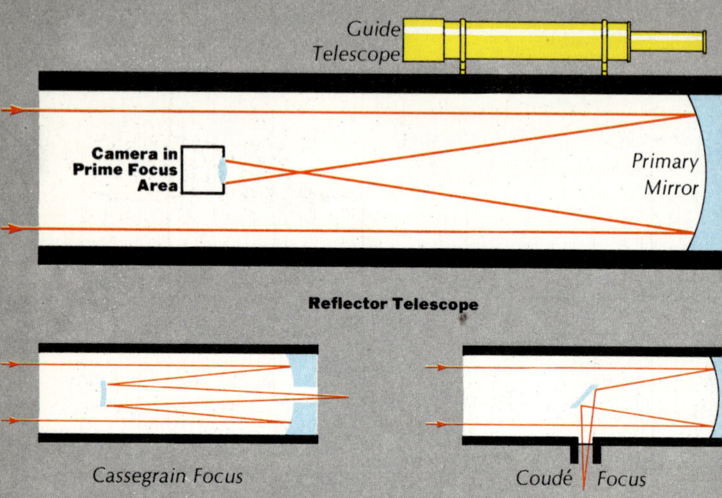

than a refractor, thereby increasing aperture and light-gathering power. There is also an increase in *resolving power*, which is the ability to separate angular distances between stars or galaxies.

In addition to the advantage of a permanent photographic record, long exposures bring out faint details not visible to the eye. The astronomer observes the sky through a *guide telescope*, which is mounted to the larger instrument taking the photographs. Although an eyepiece can be inserted at the focal point for visual observation, large refractors and reflectors usually have cameras attached at the *prime focus*. This is one of several locations possible with the versatile reflecting telescope. Using additional mirrors, the *Cassegrain focus* can reflect light back through a hole in the center of the primary mirror, thereby increasing the focal length and more conveniently locating the focal point. The *Coudé focus* permits the light to be focused down to an observing room where special equipment can be used in a controlled environment.

Radio Telescopes

Radiant energy from space—*cosmic rays, gamma rays, x-rays, ultraviolet, light, infrared* and *radio waves*—create electric and magnetic fields which propagate electromagnetic waves. The earth's atmosphere acts as a shield and permits only a fraction of this spectrum of radiation to reach the surface. The atmosphere is transparent to some ultraviolet, light, infrared, and radio waves, and is said to possess *optical* and *radio*

*Opposite: 300-foot radio telescope
of the National Radio Astronomy
Observatory, Greenbank, West Virginia*

windows. Light energy is the most familiar, since our eyes are adapted to its use. Less familiar are the other forms of electromagnetic energy, which were not understood until the present century: for example, the radio window was unknown until the advent of broadcasting. In 1931, Jansky discovered radio propagation from the Milky Way. Unlike the radio waves that are transmitted from a radio station, these radio waves cannot be heard. In radio broadcasting, sound is superimposed as a modulation of the radio wave, and this modulated wave is picked up by the receiving antenna and passed on to the receiver, where it is reinterpreted as a sound wave. The actual radio waves pass only between transmitter and receiver; the rest is electronic circuitry.

Radio waves are longer than light waves and therefore require much larger installations than optical observatories. In a radio telescope, which is analogous to a reflecting optical telescope, radio waves strike a huge paraboloid "dish," which reflects them to an antenna at the focal point. A current is induced in the antenna and amplified in a receiver, where a tuner permits the selection of the wavelengths under study. These electronic signals are recorded by a pen moving across a rotating paper drum.

The Structure of the Sun

The bright sun, which dazzles the eye, conceals the seething cauldron of activity that makes it a star. About 5 billion years ago, the sun and planets condensed from a cloud of dust and gas. Most of the cloud collapsed to form the sun with the planets no more than specks and residue of stellar evolution. As the most massive central body, the sun provides the gravitational bond keeping the earth and its companion planets in their orbital paths. The sun is immense, containing more than 99 percent of the total mass of the solar system. More than 1 million earths would be required to match the sun's mass; more than 300 earths side by side would barely ring its circumference.

The visible region of the sun is called the *photosphere,* or the light sphere. Here the energy that originated deep inside the sun bursts forth to bathe the solar system with radiation. Fortunately, the earth receives merely a fraction of the total energy emitted by the sun. At the earth's distance, the radiation from the sun can be pictured as a sphere with a radius of 93 million miles. The amount of energy intercepted by the tiny dayside of the earth is insignificant when compared with the total area of this sphere of radiation.

The sun is mainly composed of hydrogen, and, since neutral hydrogen makes the sun opaque, the photosphere can be viewed to a depth of

only a few hundred miles. This results in *limb darkening* where the sun's disk decreases in brightness toward the limb.

The nature of the interior of the sun is determined indirectly from our knowledge of the physical laws of the behavior of gases under high pressure and temperature. At the core, at a temperature of millions of degrees, the sun is converting its hydrogen to helium by *thermonuclear fusion*. In the process, some of the hydrogen is changed to radiant energy at the rate of 4 million tons per second. At the photosphere, the energy radiates into space, heating the gases to 8,000°F.

Periodically, the *rice-grained* appearance of the photosphere is interrupted with dark markings called *sunspots*. These spots, which are found in pairs of opposite magnetic polarity, are several thousand degrees cooler and seem dark against the bright photosphere. The *umbra* or central region of the spot is darker than the surrounding *penumbra*. Sunspots develop to maximum number in an average period of 11 years. This *sunspot cycle* was discovered by Schwabe in the 19th century. The cycle begins with a spot or two in the middle latitudes of the sun. Gradually more spots form closer to the solar equator until at maximum over 100 spots can be seen, with the greatest density about 15° north and south of the equator. A large spot can be many times larger than the diameter of the earth. *Faculae* are bright flame-like "little torches," seen on the darkened limb extending above the photosphere.

The *chromosphere*, or color sphere, is the first layer of the sun's atmosphere. Although it is hydrogen that makes the region appear red, other elements are also present, including helium and calcium. In fact, helium was discovered here before it was identified on the earth. The most spectacular feature of the chromosphere are the *prominences*. *Eruptive prominences* appear as huge geysers of solar matter rushing hundreds of thousands of miles into space; *quiescent prominences* appear more stable and extend as high as 30,000 miles.

Before the use of special instruments, the chromosphere and promi- **83**

Opp.: Solar flare recorded on December 19, 1973 from earth orbit aboard Skylab 4; Above: The solar spectrum identifies the composition of the sun.

Top lt.: Magnetogram of sun with strong fields associated with sunspots; Top rt.: Photosphere and solar granulations; Btm. lt.: Various regions of the sun can be studied in a single spectral line; Btm. rt.: Sunspots are cooler than brighter portions of the photosphere.

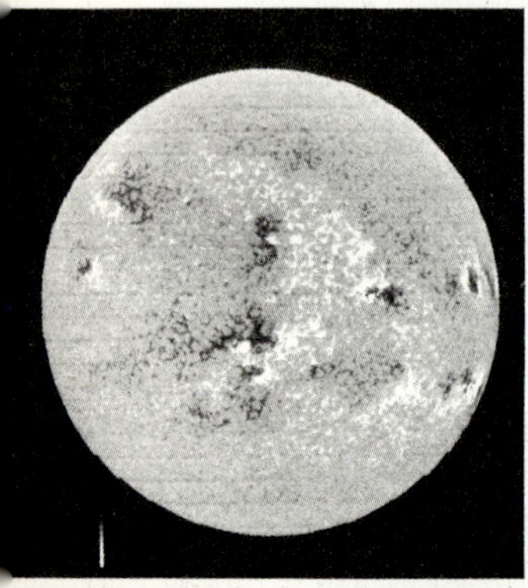

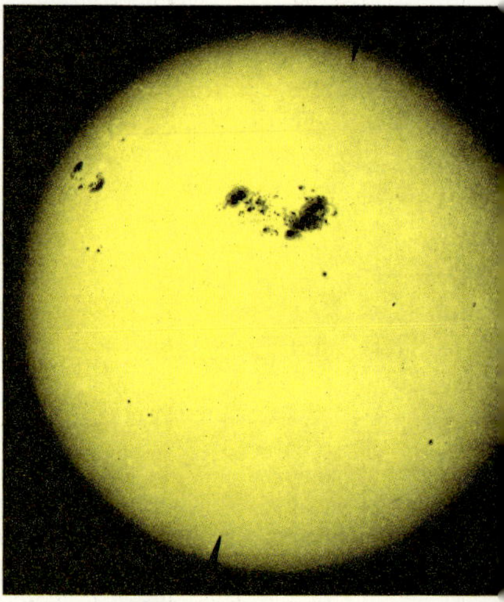

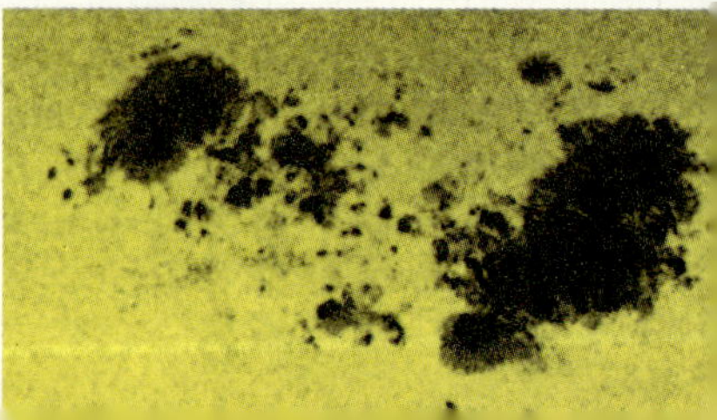

nences were visible only during a total eclipse of the sun, when the bright photosphere is covered by the moon. *Coronagraphs* occult the photosphere with baffles inside the telescope, making possible time-lapse motion pictures of prominences.

The outer atmosphere, called the *corona,* is seen during a total eclipse of the sun. The *inner corona* is gaseous and composed of solar substance. The *outer corona* is made of tiny solid particles which reflect sunlight. During an eclipse, *coronal streamers* extend radially several solar diameters.

The *radio sun* extends beyond the corona and is observed at various wavelengths with radio telescopes. Shortwave signals originate in the chromosphere; longer wavelengths are detected above the corona. Disturbances such as sunspots and *flares,* enormous outbursts from the photosphere, affect the radio sun, indicating that solar activity is inter-related and extends to the various layers of the sun.

85

Above: The solar corona, the outer atmosphere of the sun, has streamers extending millions of miles into space.

The Face of the Moon

In 1610, Galileo observed the face of the moon through a telescope. He saw *craters, mountain ranges,* and the dark lava plains called *maria,* or seas. Because the astronomical telescope inverts the image, these lunar features appear "upside down." Some of the craters (which range in size from 250 miles to less than 1 mile) have flat floors resembling the maria. Other craters are cup-shaped and are like large saucers with central mountain peaks. The larger formations are called *walled plains.* The smallest are the *craterlets.*

The maria are more prominent on the eastern hemisphere. In general, they appear circular, resembling the large craters with flat floors. Maria are rimmed by mountain ranges with a steep wall facing the maria and a gently sloping exterior wall. The largest maria, the *Oceanus Procellarum,* the *Ocean of Storms,* is irregular but seems to have been formed by overlapping circular basins.

The mountain ranges that border the maria are as high as those found on the earth. There are isolated peaks rising above the maria floor and one straight mountain range. These formations appear to be the result of lava inundation at an earlier period in lunar history. The maria were formed through successive waves of molten rock.

The telescope shows cracks, crevasses, and a valley through a mountain range. A straight wall rises above the floor like a lunar palisade. Some craters have a system of rays extending for many miles. One crater in particular, *Tycho,* has the most prominent ray system and is best seen at full moon. The craters and mountain ranges are conspicuous before and after full phase, since lunar features cast little or no shadow during full moon, making them difficult to see in the glare of reflected light.

Above: Tycho with its ray system as bright veins radiating from the crater wall; Opp.: Eastern hemisphere as seen at last quarter phase.

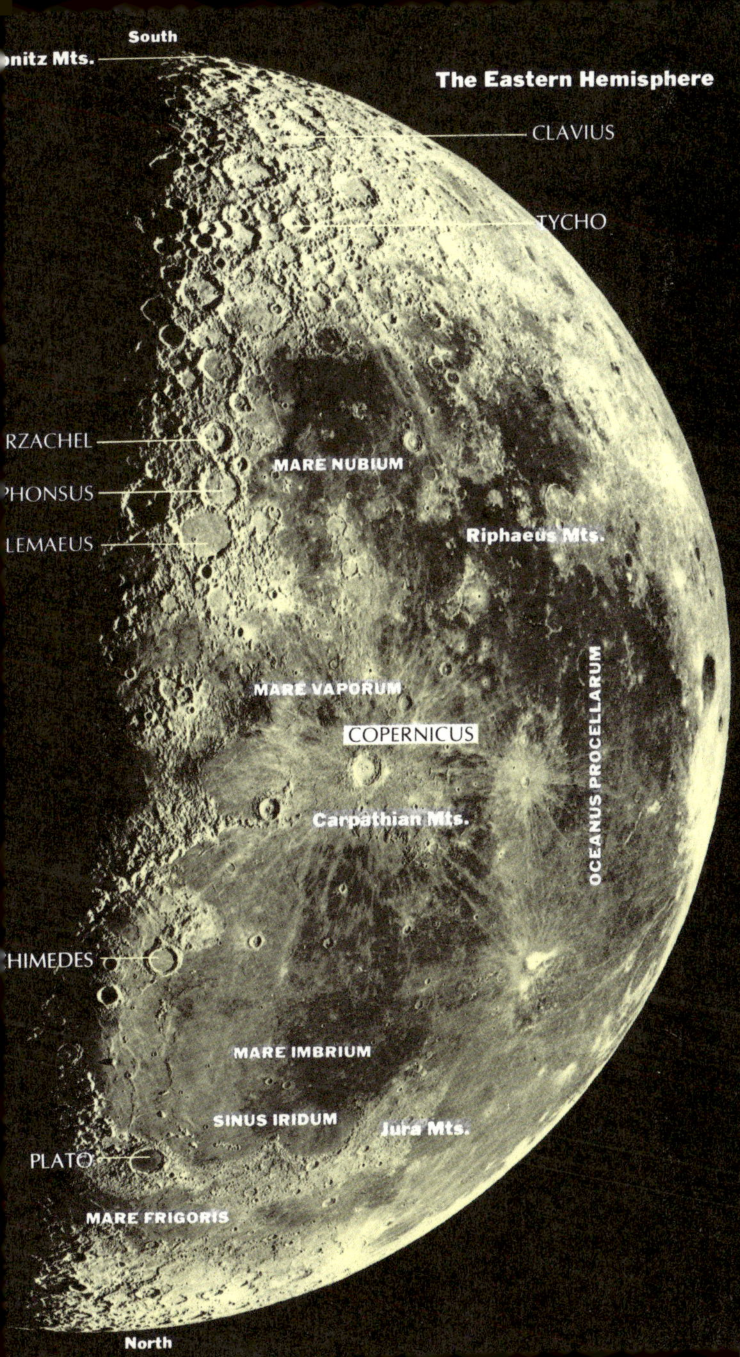
South
The Eastern Hemisphere
CLAVIUS
TYCHO
onitz Mts.
RZACHEL
PHONSUS
LEMAEUS
MARE NUBIUM
Riphaeus Mts.
MARE VAPORUM
COPERNICUS
OCEANUS PROCELLARUM
Carpathian Mts.
HIMEDES
MARE IMBRIUM
SINUS IRIDUM
Jura Mts.
PLATO
MARE FRIGORIS
North

The Western Hemisphere
Sou
Nor
Leibnitz Mts.
MARE NECTARIS
MARE TRANQUILLITATIS
MARE CRISIUM
MARE SERENITATIS
Lunar Alps
Leibnitz Mts.
Nor

The Lunar Seas

The lunar seas carry the romantic names on the moon. *Mare Imbrium*, the *Sea of Showers; Oceanus Procellarum*, the *Ocean of Storms; Mare Nubium*, the *Sea of Clouds; Mare Vaporum*, the *Sea of Vapors*, are the plains on the eastern hemisphere. On this half of the moon the maria have names pertaining to moisture. The theme is carried to smaller features of the seas such as *Sinus Iridum*, the *Bay of Rainbows*. On the western hemisphere the names reflect calm, such as *Mare Tranquillitatis*, the *Sea of Tranquillity; Mare Serenitatis*, the *Sea of Serenity; Mare Fecunditatis*, the *Sea of Fertility*. The pattern is broken by *Mare Crisium*, the *Sea of Crisis*, near the western limb.

There is little doubt that the maria were formed by magma welling up from the interior of the moon. How this came about is not fully understood. Prior to satellite exploration, astronomers suggested that maria are restricted to the nearside of the moon. Later, photographs of the farside showed a preponderance of craters. The maria were formed in successive flows at a later period in the moon's history.

Craters are seen submerged as "ghost" craters, outlined as bright rings against the dark maria background. Craters that existed before the maria surface congealed are seen in various degrees of submersion. Other craters found in the maria are complete and were formed after the lava plains were laid down. Some of these craters are surrounded by rays of bright debris. The telescope shows many smaller secondary craters surrounding the larger formations, indicating an impact origin. Cracks in the maria follow a radial pattern, suggesting that large blocks struck the moon to form these plains.

89

Opp.: The western hemisphere of the moon at first quarter phase; Above: Impact craters cover the floor of Mare Serenitatis.

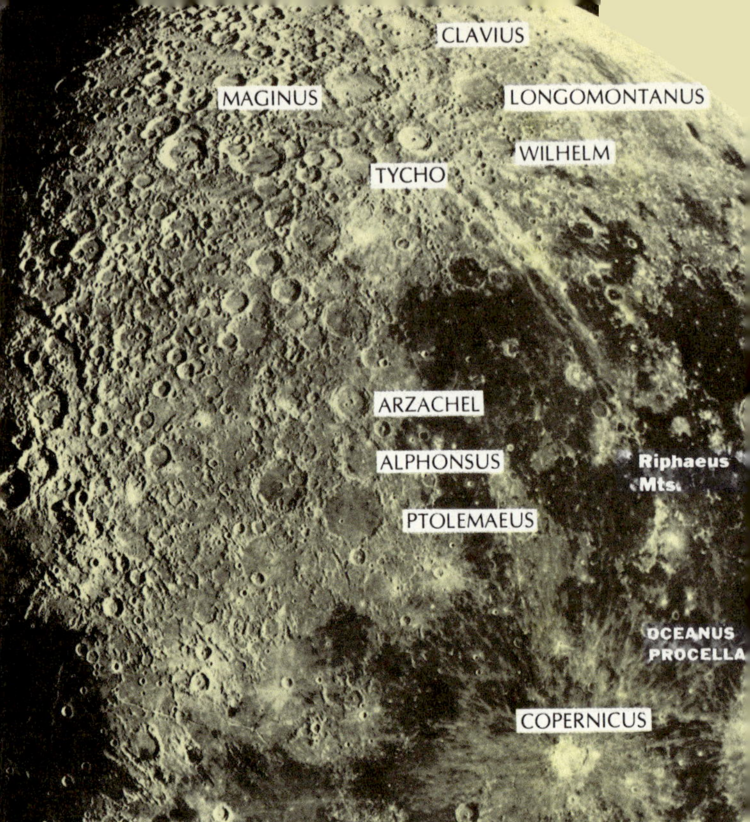

The Craters

Craters are the most impressive objects on the moon. The telescope reveals a complexity of chaotic detail especially in the highlands where craters are most numerous. Craters are named after famous scientists and philosophers, while mountains are named after ranges on the earth. *Plato* is located in the mountain range called the *Alps* on the north rim of the Mare Imbrium, and *Archimedes* is found in the Mare Imbrium near the *Apennines*. These two craters look alike, but the difference in the shading of their floors is obvious: the floor of Mare Imbrium is dark near Plato and the Alps and much lighter and at higher elevation near Archimedes and the Apennines.

Copernicus is one of the conspicuous craters toward the center of the moon in the Oceanus Procellarum. It is visible to the unaided eye as an irregularity along the terminator shortly after first quarter phase. Copernicus is interesting to watch as the terminator sweeps across the crater floor. Shadows change along the tiered interior walls. By full moon phase the sun shines almost vertically into the crater, erasing the details made visible by the contrasting shadows. Now its bright ray system is prominently displayed as a sunburst on the maria floor. Copernicus was born during the impact of a giant meteoroid crashing into the Oceanus Procellarum. The sunset terminator bathes the crater in darkness after the last quarter phase.

A conspicuous trio of craters is found along the edge of Mare Nubium. These are *Ptolemaeus*, *Alphonsus*, and *Arzachel*. Arzachel appears to be the youngest, with a sharp crater lip and a high central peak. Alphonsus is older and shows evidence of flooding. Dark markings appear along the floor at the base of the crater wall. These may be layers of volcanic ash from earlier tectonic activity. Gases have been observed exuding from the central mountain peak. Ptolemaeus, its worn walls showing signs of age and pitting by meteoroid impacts from space, may be the oldest of the three craters. Its floor is flat as a result of magma flowing to the surface from the interior, and many small craters abundantly mark the surface.

Tycho is situated in the highlands toward the lunar south pole. This region is rich in overlapping craters isolated from the maria. Tycho is unique and is believed to be one of the youngest craters. Its walls are intact and do not show the deterioration of other nearby craters such as *Maginus*, *Longomontanus*, and *Wilhelm*. The debris from the impact that caused Tycho extends as rays almost 2,000 miles across the face of the moon, over all types of terrain from Mare Nectaris to the west to the *Riphaeus Mountains* in the Oceanus Procellarum. Nearby and to the south is *Clavius*, the largest walled plain on the moon.

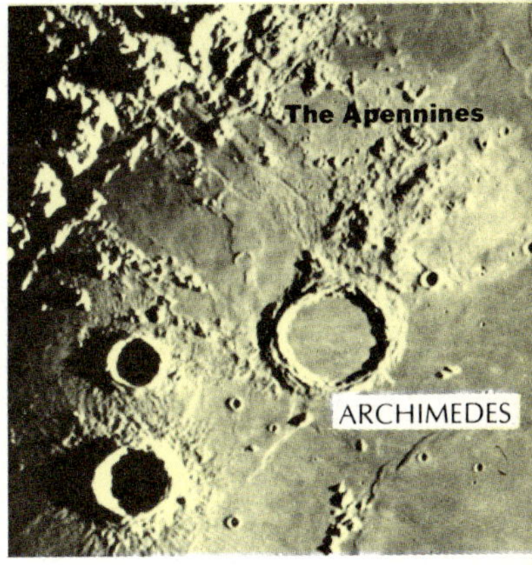

Opp.: Cratered highlands near the lunar south pole
Above: Plato in the lunar Alps and Archimedes near
the Apennines border the Mare Imbrium.

Lunar Mountains

Mountain ranges on the moon follow the circular maria lava plains. In fact, there is a strong resemblance between mountains on the moon and the walls of the large flat craters such as Plato and Archimedes. The inner face of the mountains and the crater walls have a steep slope toward the flat floor; the outside face of the crater gradually slopes away to the surrounding surface. The mountain ranges are also found to slant away gradually from the flat maria region.

The maria are believed to be formed by impact, raising a huge circular wall of an immense crater. Later, lava flows created the relatively smooth floors of the maria. Isolated ranges and peaks may be the outcrops of ancient ranges submerged by subsequent recurring flows. The Mare Imbrium has several isolated peaks such as *Pico*, *Piton*, and the *Straight Range,* a short mountain group that follows the curvature of the Mare Imbrium when connected with Pico and Piton.

Following the pattern of naming some of the mountains on the moon after ranges on the earth, the *Jura Mountains* border the Sinus Iridum. Proceeding along the rim of Mare Imbrium are the *Lunar Alps.* The prominent crater Plato, with its dark, flat floor, is imbedded in the range, apparently having impacted the Alps after the mountains were formed. Lava seeped into the crater to level its floor. There is evidence of further catastrophic events in the *Alpine Valley;* a huge cleft was formed when the mountains were literally torn apart, and lava seeped in, creating a passage between Mare Imbrium and Mare Frigoris.

The *Caucasus Mountains* along the northern rim of Mare Serenitatis curve to form the eastern edge of Mare Imbrium. Here a break in the range connects Mare Serenitatis with Mare Imbrium. South of the break are the *Apennines,* which curve southwest and terminate at the crater *Eratosthenes.* The *Doerfel* and *Leibnitz Mountains* are in the southern hemisphere. Here are the highest peaks on the moon, reaching elevations to 30,000 feet, as high as Mount Everest on the earth.

Opp.: Lunar mountains ring the maria or "seas." Above: Apollo 17 photo revealing part of the lunar farside covered with craters.

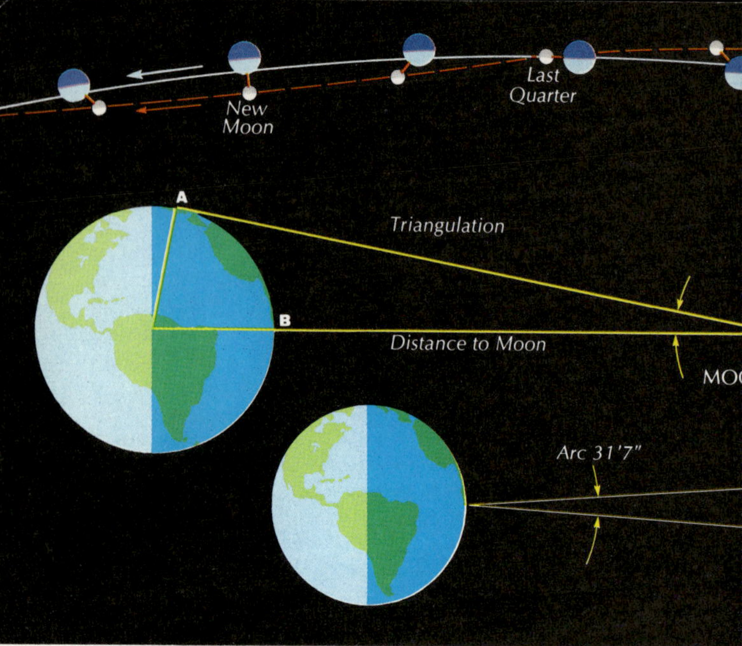

Measurements to the Moon

The distance between the earth and the moon can be determined by observing the moon from two points or stations on the earth's surface. This method, called *triangulation*, is also used by surveyors to measure distances on the earth. Since the moon is much nearer to the earth than the stars, each station will see the moon in a different star field. The amount of angular displacement or shift of the moon's position among the stars is called *parallax*. This parallactic displacement decreases as the distance to the object in space increases. For example, the stars are too far away to show a measurable parallax·from two points on the surface of the earth. Stellar parallax requires the diameter of the earth's orbit for a baseline. (See page 111.) The parallactic displacement of the moon among the stars is equal to the angle made by the two stations on the earth as seen from the moon. When the radius of the earth is the baseline the displacement is called the moon's horizontal parallax.

Parallax is used to find the diameter of the moon. Once its distance has been calculated, the moon's angular diameter can be converted to linear measure. At *perigee* (nearest to the earth) the moon will appear larger than at *apogee* (most distant from earth). The mean value of the angular diameter is 31' 7" of arc or about one half of a degree. At the moon's mean distance from the earth this angular measure is equivalent to a distance of 2,160 miles.

94

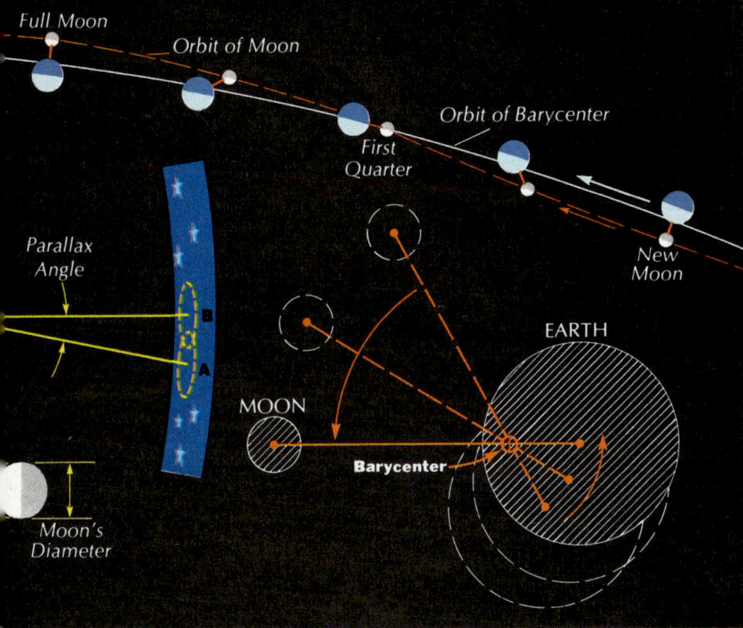

The Earth-Moon System

The gravitational effect of the moon on the earth is exemplified in the periodic rise and fall of the ocean tide. The gravitational attraction of the earth is said to keep the moon in an orbit around the earth. Strictly speaking, the earth and the moon revolve about a common center of gravity called the *barycenter*, which is located about 1,000 miles below the earth's surface.

The distance between the center of the moon and the barycenter is 81 times greater than the distance from the barycenter to the center of the earth. Since the barycenter is the center of mass, then the earth is 81 times more massive than the moon. The barycenter is located by observing the nearby planets. For example, Mars oscillates against the background stars in a period of a sidereal month. This motion of Mars is not real but the effect of the earth's center revolving about the barycenter. The amount of the displacement of Mars is a measure of the distance between the observer on the surface and the barycenter.

The earth and moon revolve about the sun in one year and about the barycenter in one sidereal month. Yet the orbits of the earth and moon remain concave to the sun. Relative to the earth, the moon revolves in an apparent elliptical orbit with the earth at one focal point. Relative to the sun, the earth and moon revolve around the barycenter, which revolves in an apparent elliptical orbit around the sun.

The Solar System

The Planets

The planets may be divided into two groups, the *terrestrial* or earth-like and the *Jovian* or Jupiter-like planets. Interestingly, the physical and orbital characteristics of one group are the opposite of those of the other. The terrestrial planets include *Mercury, Venus, Earth,* and *Mars*. The Jovian planets are *Jupiter, Saturn, Uranus,* and *Neptune. Pluto,* though terrestrial in size, is the remotest planet in the solar system.

The terrestrial planets are near the sun. Consider the earth-sun distance of 93 million miles as the yardstick or unit of measure. This distance is called the *astronomical unit,* or A.U. Mars revolves in an orbit at 1.5 A.U. from the sun, or half again as distant as the earth. Jupiter

lies over 5 A.U. from the sun. The wide gap between Mars and Jupiter, the boundary of the terrestrial and Jovian planets, is filled with thousands of blocks of rock and iron called the asteroids, the *minor planets*. With Jupiter at 5.2 A.U. and Neptune at 30 A.U., the Jovian planets are found to be deep in the solar system, with many astronomical units between them.

The terrestrial planets rotate slowly, taking between 100 and 250 days to complete one turn; the Jovian planets spin in less than one day. Earth-like pl nets take days to revolve around the sun; Jovian planets require year to complete their orbital journeys. Terrestrial planets are small and rocky; Jovian planets are huge and composed of gaseous elements. All the planets revolve in elliptical orbits so that their distances from the sun are continuously changing.

Mercury

Mercury, with a diameter of 3,000 miles, is the smallest as well as the planet nearest the sun. Since its orbit is very eccentric, the planet is almost 15 million miles nearer the sun at perihelion than at aphelion. Its mean distance from the sun is 36 million miles or 0.39 A.U. At a mean orbital speed of 30 miles per second, the planet requires 88 days to complete its journey around the sun.

As Mercury revolves, from the earth the planet appears to alternate from east to west of the sun. The angle between the planet and sun is called *elongation*. Because of its proximity to the sun, Mercury's elongation cannot exceed 28°, appearing low on the horizon at sunrise and sunset. From earth, the telescope reveals Mercury passing through phases like those of the moon, as various parts of the planet's dayside (which faces the sun) are exposed to the earth during the planet's revolution around the sun.

Mercury's size and proximity to the sun makes surface observation difficult. Dark and light markings have been seen and these appear to be flat maria regions like those found on the moon. Radar measurements confirm the irregularity of the surface.

For years it was believed that Mercury's period of rotation was equal to its revolution of 88 days. Radar observations made in 1965 and later photographic confirmation showed that the planet rotates in about 59 days or two-thirds of the revolutionary period. A combination of these motions makes the days and nights on Mercury each 88 earth-days in duration. At noon, surface temperatures rise to 640°F, or 337°C.

Venus

Venus is the planet nearest the earth and one of the least understood. Perpetually shrouded in cloud cover so that no surface details are visible, through the telescope Venus appears in different phases from crescent to full. Galileo was the first to see the phasing as one proof of the Copernican heliocentric system. Venus is about 7,700 miles in diameter and revolves at a mean distance of 67 million miles from the sun. Its mean orbital speed is 22 miles per second, which is about eight miles per second slower than Mercury. At a greater distance from the sun, Venus needs less speed to maintain its orbit in a period of 225 days.

The cloud cover makes the planet a disappointment when viewed through the telescope. Occasionally, dark areas have appeared among the clouds and unsuccessful attempts have been made to measure the period of rotation by the passage of these spots across the disk of the planet. An ashen glow in the atmosphere on the nightside suggests the refraction of light from the bright hemisphere facing the sun. Electric discharges and aurora may also be present. About 40 years ago, carbon

dioxide was found to be the most abundant substance in the atmosphere. Since then, the nature of the clouds and the presence of water
vapor has been investigated. High altitude studies have been made for
15 years in search of oxygen and nitrogen, which constitute the atmosphere of earth. The nature of the clouds is still under investigation.

Radar signals reflected from the surface of Venus indicate a rough
surface and possibly a mountain range and ravine in the southern
hemisphere. Signals indicate a rotational period of 243 days
retrograde—in the reverse direction of the other planets. Beneath the
clouds the temperature is high and, prior to satellite investigation, was
estimated to be more than 212°F, or 100°C, the boiling point of water.

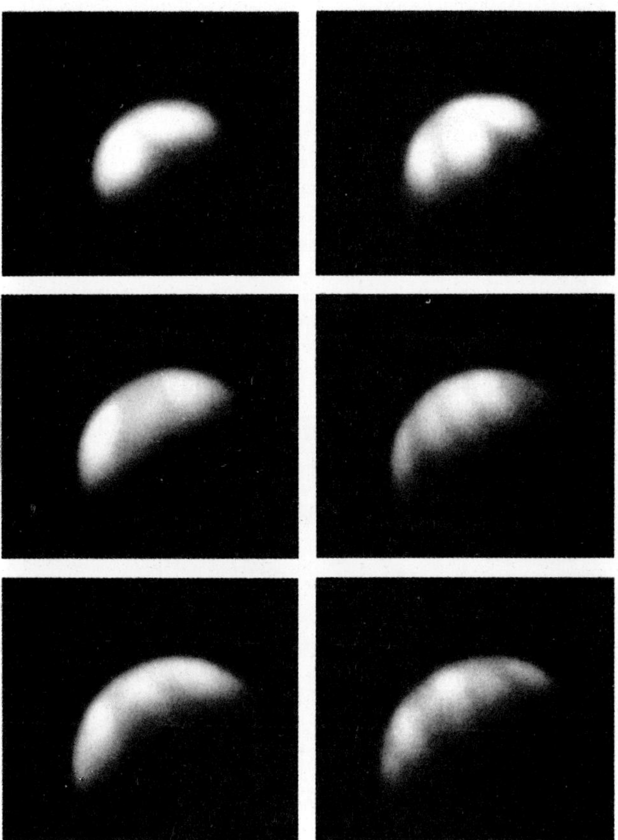

*Venus is seen in crescent
phase as the planet passes between
the sun and the earth.*

Earth

The earth is spheroidal with slight polar flattening due to rotation. It is the third planet from the sun and is the only one with extensive bodies of water which cover more than 70 percent of the surface. A thin portion of the crust extends above the oceans. These are the continents, which, together with the oceans and part of the atmosphere, constitute the *biosphere* where life can exist. The solid crust called the *lithosphere* is only between 3 and 40 miles thick and is separated from the next layer, the *mantle,* by the *Mohorovičić Discontinuity.* Here a sharp change in structure takes place. The mantle extends 1,800 miles toward the center and, although not liquid, movement of the rock does take place due to the great pressures that exist. Below the mantle is another discontinuity, which is the boundary of the *core.* The core is believed to be divided into a liquid layer, 1,350 miles thick surrounding a solid center about 1,600 miles in diameter. Both the liquid and solid portions of the core are believed to be iron and nickel, which acts as a dynamo as the earth spins, creating a magnetic field extending thousands of miles into space.

The earth's *atmosphere* is mainly nitrogen (78 percent) and oxygen (20 percent) with an abundance of water vapor that condenses into clouds. This mixture extends beyond the *stratosphere* to the *mesosphere,* 55 miles above the surface. Here the *ionosphere,* or the electrified portion, of the atmosphere begins, with its various ionized layers at different altitudes. Theoretically, the atmosphere extends to 22,000 miles. Most of the air, however, lies below 4 miles altitude and decreases rapidly in density and pressure with increase in elevation. Outer space begins 100 miles above the earth.

Mars

Through a telescope, the disk of Mars at favorable opposition shows as much detail as a naked-eye view of the full moon. Since the planet's atmosphere of carbon dioxide is transparent, the surface shows polar caps of frozen carbon dioxide, ocher-colored, desert-like expanses, and grey-green maria regions once thought to be seas. No large bodies of water exist.

Hydrated iron oxide is assumed to account for the reddish hue, although it might be attributed to carbon suboxide. The greenish areas become pronounced during spring and summer of the Martian year, as the polar ice cap recedes from the middle latitudes. Some astronomers have attempted to explain the green areas as Martian vegetation, but studies have failed to reveal the presence of chlorophyll, making vegetable life highly unlikely.

The atmosphere is predominantly carbon dioxide, with traces of water vapor, oxygen, and ozone. Atmospheric pressure is about one-hundredth that of the earth. Blue clouds of ice crystals form miles above the surface, while yellow clouds of dust are churned by raging winds.

Mars has a diameter of 4,200 miles and is one-tenth as massive as the earth. The period of rotation is 24 hours and 37 minutes; the equator is inclined 25° to the orbit. Temperatures can reach 40° at the equator but fall below −100°F at night.

Mars has two satellites, discovered by Asaph Hall in 1877. Only a few miles across, they are assumed to be captured asteroids with Phobos 3,700 miles above Mars revolving in 7 hours and 39 minutes and Deimos revolving in 30 hours and 18 minutes at a distance of 12,500 miles. **101**

Asteroids

The *asteroids* are small bodies that in general occupy orbits between Mars and Jupiter. The largest, *Ceres*, is 480 miles in diameter and was discovered in 1801. There is a wide gap in space between Mars and Jupiter. By applying a mathematical progression called *Titius'-Bode's law*, astronomers predicted that an undiscovered planet occupied the gap. Ceres was hailed as the missing planet. Shortly after Ceres' discovery other asteroids such as *Pallas, Juno,* and *Vesta* were found. Today, thousands of these *minor planets* are known, suggesting that at one time two or more proto-planets collided, forming the *asteroid belt*. But acceptance of this theory of cataclysmic origin is not necessary if the asteroids are the remains of the original solar nebula of insufficient mass to form a *major planet.*

Not all asteroids remain in orbits between Mars and Jupiter. *Icarus* passes from the asteroid belt to within 20 million miles of the sun, inside the orbit of Mercury. Earth-grazers such as *Eros, Amor,* and *Apollo* approach within 14 to 10 million miles of our planet. In 1937, *Hermes* was discovered at a distance of about 500,000 miles or twice the distance to the moon. Can the earth collide with an asteroid? The surfaces of the moon, earth, and other planets confess to earlier collisions with asteroid type bodies. Meteoroids or large blocks of rock and metal weighing several tons continue to strike the earth periodically but none attains the size of a large asteroid capable of catastrophic damage, presumably because the larger blocks were already used up in creating the planets. Since observation of the moon began (with Galileo), no craters—large or small—have been added to its surface.

Jupiter

Jupiter is the largest planet in size and mass. In fact, Jupiter contains more matter than all the planets, satellites, asteroids, dust, and gas that make up the rest of the solar system. Together, the sun and Jupiter constitute 99.9 percent of the solar system. Jupiter's equatorial diameter is 89,000 miles, while its polar diameter is almost 84,000 miles.

The planet is an interesting telescopic object. Jupiter's atmosphere is marked with horizontal bands: dark bands are referred to as *belts;* bright bands are called *zones.* The *Great Red Spot* is an unusual atmospheric feature. The spot is elliptical, with a major axis of 24,000 miles; its width is almost 8,000 miles.

Like the sun, Jupiter is mainly hydrogen and helium. The planet's atmosphere may be 85 percent hydrogen and 15 percent helium by volume. Included are traces of ammonia and methane, with ammonia crystals in the upper atmosphere. The atmosphere is believed to be a few hundred miles thick and the consistency of wet slush at the surface. Jupiter's surface may be liquid hydrogen with solid hydrogen at the center, or there may be a rocky central core.

Four satellites, the *Galilean moons* discovered by Galileo, are visible through binoculars or a small telescope. These interesting objects may be followed from night to night changing positions as they revolve around Jupiter. They are named *Io, Europa, Ganymede,* and *Callisto.* Ganymede is larger than the planet Mercury. The other three are comparable in size to the earth's moon. Eight other satellites—for a total of 12—that revolve about Jupiter are only a few miles across. Four of these may be captured asteroids.

Opp.: An asteroid is detected as a short streak of light among the stars. Above: Jupiter, with its banded atmosphere and Great Red Spot.

Saturn

Saturn is the loveliest planet to observe in the telescope. Faint markings, similar to Jupiter's, that appear across the disk are clouds of hydrogen and methane. Saturn is believed to have a surface of liquid hydrogen which merges into solid hydrogen in the interior. The planet rotates rapidly on its axis, giving it a pronounced equatorial bulge. Its diameter is 75,000 miles at the equator and 68,000 miles through the poles.

But the striking feature of Saturn is its rings, composed of millions of particles which revolve in the plane of the planet's equator. The *outer ring* has a diameter of 171,000 miles. Between the outer and *middle ring* there is a separation of about 1,800 miles called *Cassini's division*. The gap is caused by the gravitational effect of Saturn and its satellites that lie beyond the ring system. Inside the bright middle ring is the faint *crepe ring* which is so tenuous that starlight can be seen shining through. A fourth ring, nearest to Saturn, is more elusive than the crepe ring and can be observed only in large telescopes. The ring system is inclined 28° to the ecliptic. Since the planet requires 29½ years to revolve around the sun, the rings are seen open and on edge every 7½ years. The rings face the earth every 15 years, alternately showing the northern and southern hemisphere. When the rings are seen edge-on, they disappear, indicating a thickness of merely a few miles.

Saturn has 10 satellites; *Titan,* the largest in the solar system, is about 50 miles larger in diameter than Jupiter's Ganymede. It is even larger than the planet Mercury.

The Twin Giants—Uranus and Neptune

Uranus was discovered by William Herschel in 1781. The planet is a sixth-magnitude object that was observed prior to Herschel but was marked on sky maps as a faint star. After its discovery, Uranus was identified on these maps providing information about its orbital motion. The planet is about 29,000 miles in diameter and twice as far from the sun as Saturn. At that distance, Uranus requires 84 years to revolve around the sun. The orbit of Uranus is inclined only 0°.46 to the ecliptic; yet its axis is inclined no less than 98° to the perpendicular of the orbit, with the north pole of the planet 8° below the orbital plane. This extreme inclination places the sun in the zenith everywhere on the planet sometime during its 84-year period of revolution. In a small telescope Uranus appears as a tiny green disk and therefore is not as interesting to observe as the planets nearer to the earth. Like other giant planets, Uranus has an abundance of hydrogen, and methane has also been discovered in the planet's atmosphere. Uranus has five satellites, the largest, called *Titania,* is about 600 miles in diameter.

Neptune is half again more distant in space than Uranus. Neptune's discovery in 1846 was a triumph for Newton's gravitational theory. When the motion of Uranus did not conform to prediction, some astronomers suggested that the law of gravitation was not universal. The gravitational theory was confirmed by two astronomers, Leverrier in France and Adams in England, who accounted for these errors in the position of Uranus by the effect of an unknown planet more remote from the sun. Thus Neptune was the first planet to be predicted and confirmed from mathematical computations. At a distance of 30 A.U., Neptune is difficult to observe even in a large telescope.

Neptune and Uranus are the giant twins in the solar system, with similar diameters and atmospheres of hydrogen and methane. Neptune has two satellites, *Triton* and *Nereid;* Triton is about 2,400 miles in diameter—about 300 miles larger than the earth's moon.

Opp.: Saturn; the rings may be the remains of a satellite. Above: left, Neptune; right, Uranus; the twin giants.

Pluto

The discovery of Neptune was a triumph for the application of Newton's laws to the mass and motions of the planets. The search for other planets culminated in the discovery of *Pluto* by Tombaugh in 1930. The *perturbations* or deviations from predicted positions of Uranus and Neptune led to the search for Planet X. In 1915, Lowell calculated where a trans-Neptunian planet should be found. Fifteen years later, Pluto was discovered on photographs taken at separate intervals of time. When the plates were compared, Pluto appeared as a small fifteenth-magnitude point of light displaced among the stars by its orbital motion.

There are several characteristics that make Pluto unique. With an orbital eccentricity greater than that of any other planet, Pluto at perihelion is within the orbit of Neptune. The plane of the orbit, however, is inclined 17° to the ecliptic and there is little likelihood that Pluto and Neptune will ever collide. Pluto reaches perihelion in 1989 and will be within the orbit of Neptune for the remainder of this century.

Other Members of the Sun's Family

In addition to planets, satellites, and asteroids, the solar system contains comets, meteoroids, dust, and gas. Comets originate deep in space, perhaps in a cloud of comets extending several light-years from the sun. Between the stars and the sun, comets are small spheres of ice. Perturbations drive these comets inward to the sun where the gravitational attraction of the planets changes their orbits into elliptical curves about the sun. The most famous *periodic comet* is *Halley's Comet,* which approaches the sun every 75 years.

As a comet nears the sun, the outer layer vaporizes and forms a gaseous envelope called a *coma* around the solid *nucleus,* which contains water ice. Beyond the coma is a cloud of hydrogen gas. Particles of iron and rock trapped in the ice nucleus are released in the coma. Solar radiation exerts a pressure on the coma and forces the particles and gases away from the direction of the sun, forming a tail. A *solar wind* of high energy particles ionizes the gases in the tail and makes them bright. Particles in the tail shine by reflected sunlight.

Billions of particles are left in the comet's wake to add to the supply of interplanetary dust. As the earth revolves, it sweeps up these particles which plunge through the atmosphere to burn out as bright streaks called *meteors. Sporadic meteors* can be observed any clear night of the year. *Periodic meteors* appear in *showers* from a particular point in the sky called a *radiant.* These meteor showers are named after the constellation in the direction of the radiant. Periodic meteors are believed to be caused by the particles left by comets, and can be predicted from the known intersection of the cometary particles with the earth's orbit.

Top: Pluto, detected by its motion among the stars
Mid.: Halley's Comet (will reappear in 1986); Btm.: Halley's Comet
observed in 1066 shown on Bayeux Tapestry.

ISTI MIRANT STELLA

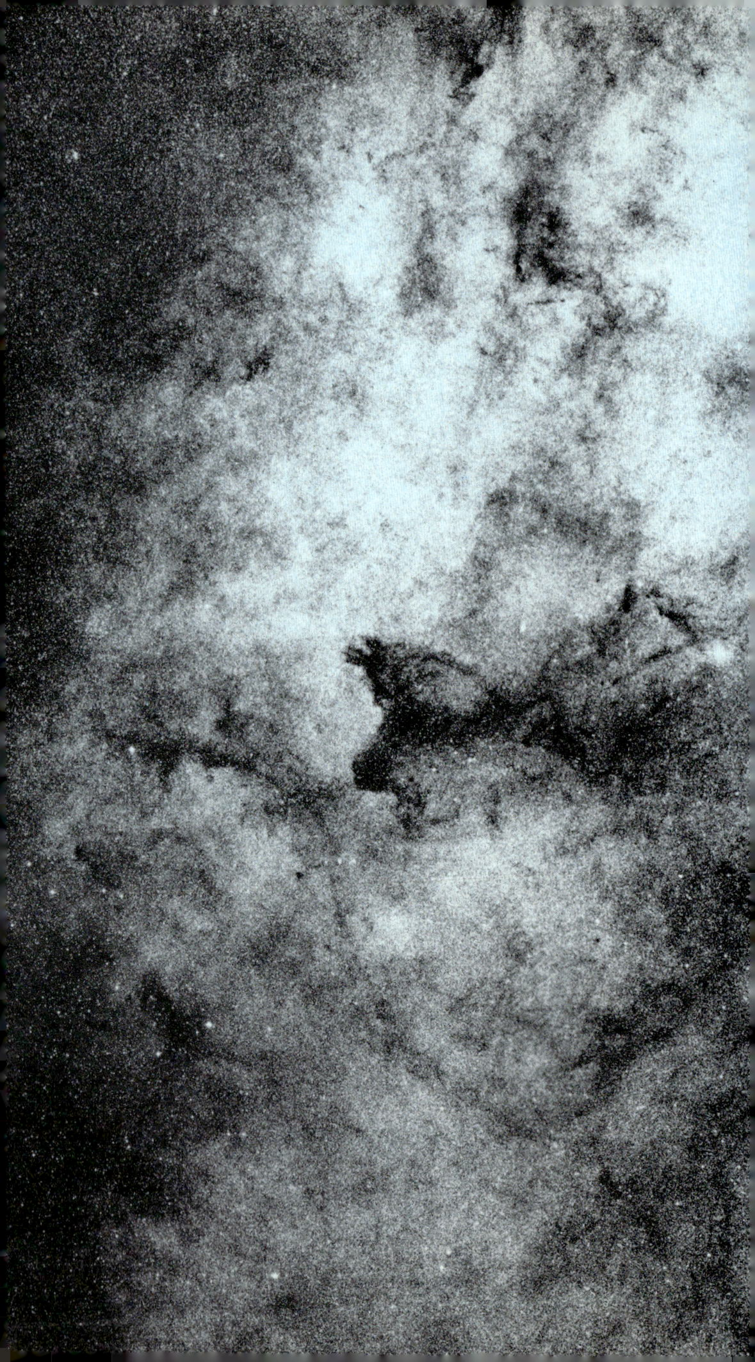

Stars and Their Evolution

Brightness and Magnitude

Magnitude is the measure of the brightness of a star. This comparison scale dates back to Hipparchus (190–125 B.C.), a Greek astronomer who grouped the stars in his catalog into six categories. The brightest stars were classified in the first group and were called first-magnitude stars. The sixth group containing the faintest stars visible to the unaided eye were the sixth-magnitude stars. The system was perpetuated by Ptolemy in 140 A.D. in his great work, the *Almagest*, which was a collection of the astronomical work of the past. Today, this magnitude scale is used with refinements made possible by modern instruments, and the present system was established in the middle of the last century by *Pogson*. *Pogson's scale* retains the old magnitude system and assigns more exact magnitudes to the stars.

One hundred times more light energy is received from a first-magnitude star than from a sixth. Since the difference in magnitude is five and the ratio in brightness is 100:1, each magnitude represents a change in brightness by a factor of about 2.5. A first-magnitude star is 2.5 times brighter than a second. Second-magnitude stars are 2.5 times brighter than third. The brightest star, Sirius, has a magnitude of −1.4 which is brighter than first magnitude. On Pogson's scale, objects brighter than first will be zero (0) magnitude. Objects brighter than zero have minus (−) magnitudes. Magnitudes increase numerically as brightness decreases. The sun's magnitude is −26.5. The faintest star photographed is +23.5. These magnitudes describe the brightness of stars as they *appear* to the eye or on a photographic plate. This *apparent magnitude* is not a measure of the intrinsic or real brightness which can be found when the distance to the star is known.

Opp.: The Sagittarius region of the Milky Way is rich in star clouds, dust, and gases. Above: A longer photographic exposure of the same field reveals more stars.

Color

Stars are incandescent globes of gases at high temperatures. Energy generated in the interior emerges at the visible layer, called the *photosphere,* and radiates into space. Other invisible wavelengths are emitted including infrared, radio, ultraviolet and x-rays. Although stars radiate energy in all wavelengths, their surface temperatures differ by tens of thousands of degrees. Stars similar to the sun have temperatures between 4,000° and 6,000° Celsius and appear yellow in color. The hottest stars radiate at temperatures from 40,000° to 100,000°C and are blue in color. Red stars are the coolest with temperatures as low as 2,000°C. The colors of the stars indicate their temperatures. Although a red star has a very low temperature compared with a blue, all stars are hot enough to vaporize all substances including the metals. In blue stars, helium, oxygen, and nitrogen atoms are *ionized.* Low-temperature red stars are inactive, allowing atoms to remain neutral and to form molecules such as titanium oxide.

Spectroscopy, the analysis of the spectrum, provides much information, including the temperature and chemical composition of stars. When starlight is separated into the colors of the spectrum, bright and dark lines are observed which relate the star's emission and absorption of energy, its axial rotation and space motion, as well as its chemical composition, temperature, mass, and diameter. Stars are classified according to their *spectral type* and are given an identifying letter. High-temperature blue stars are O-type stars. Red stars are M-type. In order of decreasing temperature, the spectral types are O-B-A-F-G-K-M. The sun, a G-type, is a star of average temperature.

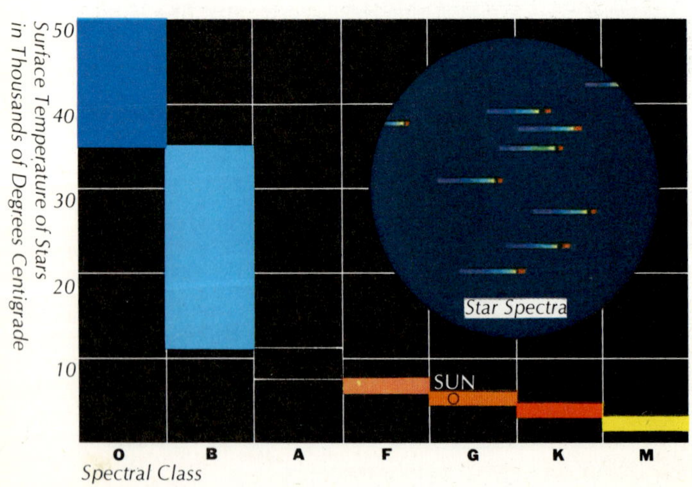

Distance

The stars are at such vast distances that even the nearest beyond the sun cannot be resolved into a disk by the most powerful telescopes. In the solar system, distances are measured in millions of miles or in astronomical units. To the stars, distances are so great that measurement in miles or even astronomical units becomes as practical as measuring the circumference of the earth in inches and centimeters. The distance to the moon was found using *lunar parallax,* the angular displacement of the moon in the sky when observed from two stations on the earth. (See page 94.) In a similar manner, the distance to a star can be found with *stellar parallax.*

Unfortunately, the stars are so far away that the angle made from two points on the earth is entirely too small to be measured. In order to cause a displacement in the position of the nearby stars, sightings must be made from opposite sides of the earth's orbit. Stellar parallax is the angle to the star made by the astronomical unit, the mean distance between the earth and sun. There are 3,600 seconds in one degree, yet no star is near enough to have a parallax of as much as one second of arc. One second of arc is the angular separation between the earth and sun when viewed at a distance of about 206,265 astronomical units or *one parsec* (a *parallax of one second).* There are 3.2 *light-years* in one parsec. A light-year is the distance light travels in one year. The nearest star, Rigil Kentaurus (Alpha Centauri) has a parallax of 0".76 and a distance of 1.3 parsecs or 4.3 light-years. At about 30 parsecs, parallax becomes small and difficult to measure, so more distant stars are measured by other methods.

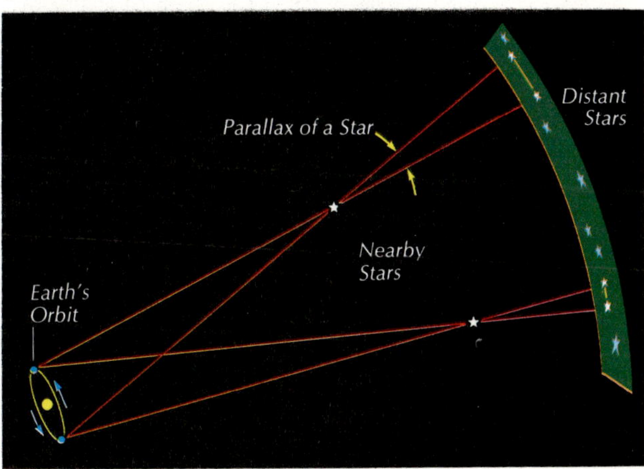

Absolute Magnitude

Apparent magnitude does not represent the intrinsic brightness of a star. A nearby star of low luminosity can appear brighter than a high luminosity star at a great distance. If all the stars were at the same distance from the earth, a comparison of their apparent magnitudes would be a measure of their "real" or intrinsic brightness. This can be accomplished by considering all of the stars at a standard distance. The *absolute magnitude* of a star is the apparent magnitude it would have at a distance of 10 parsecs. Since most stars are more remote than 10 parsecs, absolute magnitude is usually numerically smaller than apparent magnitude. If Rigel (in Orion), for example, were as close as this, the star would be 600 times brighter. Sirius decreases in brightness since its true distance is only 2.7 parsecs from the earth. If the sun were transported 10 parsecs it would appear as a fifth-magnitude star hardly visible to the eye. A comparison between the sun and Rigel staggers the imagination. Rigel at 10 parsecs has a magnitude of -6.8. With the sun at $+5$ magnitude, the difference is 11.8 magnitudes, or over 50,000 times in brightness.

Another interesting comparison is made between Deneb and Altair, two of the stars in the Summer Triangle. The apparent magnitudes of these stars is about the same with $+0.77$ for Altair and $+1.26$ for Deneb. The colors of the stars are about the same, too. If apparent magnitude were the only concern, these two stars are alike; but a measurement of parallax and absolute magnitude shows the differences. Altair is a nearby star only 5 parsecs away or half the way to the standard distance of 10 parsecs. Deneb is almost 500 parsecs distant or 100 times deeper in space. Accordingly, Deneb is nearly 10,000 times brighter than Altair. Since their colors and apparent magnitudes are about the same, Deneb must be huge—many times greater in diameter than Altair. One can appreciate Deneb when this star is compared to the sun, which is smaller than Altair.

Earlier, color was found to be related to temperature. A blue star is very hot while a red star is cool. At 10 parsecs, blue stars are found to be brighter than red stars of the same population or type. These are the *blue giants* and the *red dwarf* stars. For example, although Rigel, a blue giant, is more than 50,000 times brighter than the sun, it is 500 million times brighter than a red dwarf star called *Proxima Centauri*. This dwarf is a member of the Alpha Centauri triple star system and at the present time the nearest star to the sun. Betelgeuse, the red star in Orion, is about the same color as Proxima Centauri. Yet, at 10 parsecs, Betelgeuse is only a magnitude fainter than Rigel. Betelgeuse is enormous in size—a *red giant* many times larger than the blue giant stars. The red giants are

 among the most luminous stars known.

Diagraming the Stars

Once the color or temperature and absolute magnitudes are known, the similarities and differences of the stars can be studied. Previously, Deneb was compared with Altair and Betelgeuse with Proxima Centauri. Rigel was found to be an extremely luminous star. The sun seemed to be average in temperature and brightness. To gain a complete understanding, many stars must be investigated and classified.

At the beginning of this century, two astronomers, Hertzsprung of Denmark and Russell of the United States independently made comparison studies of stars. Their results are graphically portrayed in the so-called Hertzsprung-Russell diagram, a graph of a star's temperature or color with respect to its absolute magnitude or luminosity. The color is designated by the spectral class of the star measured along the horizontal axis (abscissa) of the graph. The highest temperature is to the left (spectral type O stars), while the lowest is to the right (spectral type M stars). The vertical axis (ordinate) contains the absolute magnitude measurement, with the most luminous stars having minus magnitudes at the top and decreasing to the plus magnitudes at the bottom of the scale. A yellow star similar to the sun, spectral type G2 and +4.8 absolute magnitude will fall in the center of the graph. Rigel, spectral type B8 and −6.8 absolute will be located in the upper left. Proxima Centauri, a spectral type M5 and +15 absolute, is found in the lower right corner of the diagram.

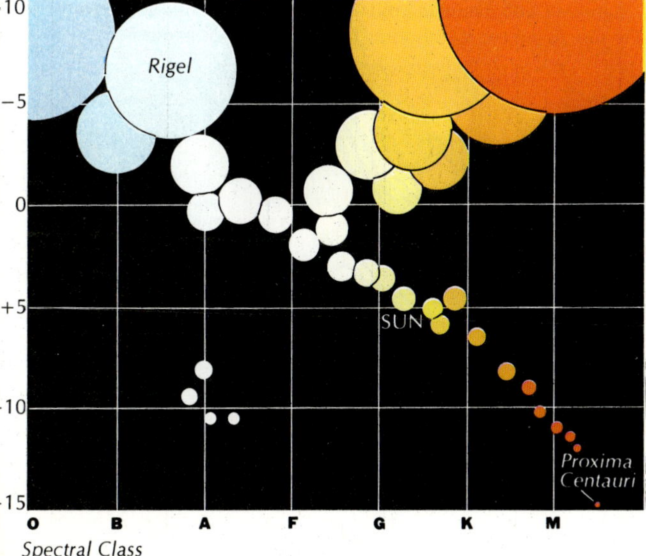

Nearest to the Sun

Assuming that the sun is located in a typical region of the star system, a sampling of the sun's neighborhood will give information about the kinds of stars that are found there as well as their abundance. Within a radius of 5 parsecs from the sun, there are about 60 stars. If these stars are plotted on the Hertzsprung-Russell diagram, they form a diagonal line from the center of the diagram to the lower right with most of the stars of type M. Of these stars, three, including Alpha Centauri A, Sirius, and Procyon, are higher on the diagram and therefore hotter and more luminous than the sun. Alpha Centauri A is the brightest star of the three that make up the Alpha Centauri triple star system. The B and C component (Proxima) are fainter than the sun. More than 50 percent of the nearby stars are M-type, with high numerical magnitudes between +10 and +15. These stars are cool and faint and are therefore small, not much larger than the planet Jupiter. They are called *red dwarfs.*

Evidently most of the stars in the Milky Way are small, with the sun a substantial member in its neighborhood. Alpha Centauri A is a yellow G-star, Procyon is type F and creamy-white, while Sirius is a white type-A star. Sirius and Procyon have strange companion stars called *white dwarfs,* which are type-A stars as faint as and smaller than the red dwarfs. Sirius, a type-A star, is one and one-half the sun's diameter and twice as massive. Its white dwarf companion has as much mass as the sun but is about the size of the earth. On the *main sequence* (as the diagonal line across the H-R diagram is called), stars diminish in mass, magnitude, and diameter from blue-white Sirius, half again as large as the sun, to the red dwarfs, only one-tenth the size of the sun.

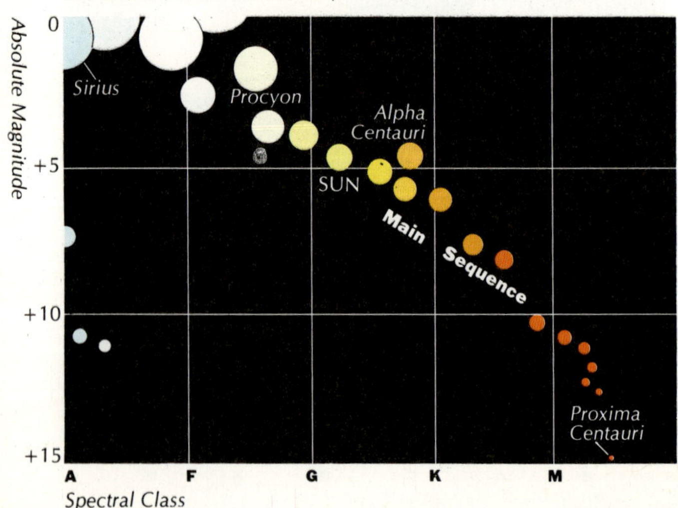

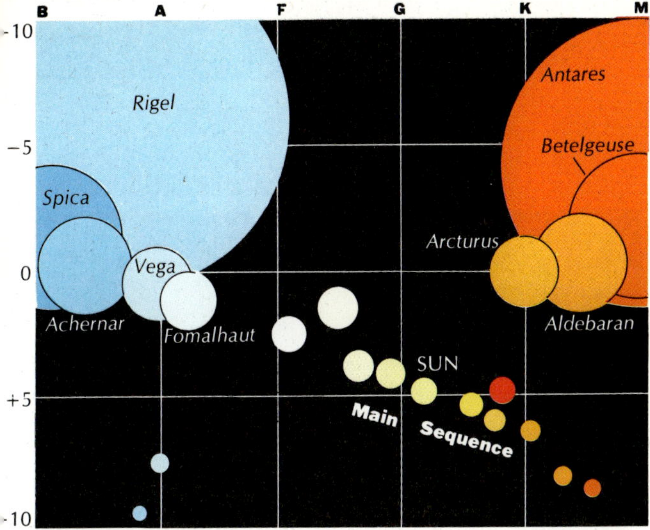

Brightest Stars

The red dwarfs with +10 to +15 absolute magnitude are too faint to be observed without optical aid. Yet there are several bright-red stars, including Betelgeuse and Antares, that are too distant to be neighbors of the sun. There are orange stars, Aldebaran and Arcturus, that are also bright as well as very distant. The orange stars near the sun are faint and on the threshold of vision. Only three of the 20 brightest stars are neighbors of the sun.

If the 20 brightest stars were placed on the H-R diagram, half would fall on the main sequence above Sirius completing the population along the diagonal line. These main sequence stars include Vega, Achernar, Spica, and Fomalhaut. The blue giant Rigel is slightly to the right and off the main sequence. The stars higher on the diagram than Sirius are more massive, hotter, and larger in diameter. From the blue giants to red dwarfs the alignment of the stars is remindful of a string of multicolored, sparkling beads.

The remainder of the brightest stars are not members of the main sequence. Betelgeuse is red but also one of the most luminous stars with a placement on the H-R diagram to the right above the main sequence. Betelgeuse is a *red supergiant,* a highly luminous star with an enormous radiating surface. Orange giants, Arcturus and Aldebaran, fall between the red giants and the main sequence. There are pulsating, eruptive, and exploding stars that occupy distinctive positions on the diagram. **115**

Clusters of Stars

There are two general classes of star clusters, *galactic* or *open clusters* and *globular clusters*. Galactic clusters are found in the plane of the Milky Way. The *Pleiades* and *Hyades* in Taurus are well known and familiar open clusters. In the southern hemisphere, the *Jewel Box* in Crux is a famous cluster. Galactic clusters are sparsely populated and lack central condensation. The Pleiades are among the best known, with six stars visible without optical aid and hundreds that can be resolved in a rich field telescope. These stars are similar to the sun and others on the main sequence.

Globular clusters are different in population as well as location in the Galaxy. These clusters contain stars that increase and decrease in brightness in less than a day. They are called RR Lyrae stars and are used to measure the distances to the clusters. In addition there are yellow, orange, and red giant stars. Generally, globular clusters form a halo around the center of the Galaxy. It was this distribution of clusters and the presence of RR Lyrae stars that located the nucleus, size, and structure of the Galaxy as well as the position of the sun in the spiral arms. The stars of globular clusters, believed to be among the oldest in the Galaxy, are highly concentrated toward the center and number in the tens of thousands.

116

Top: M13, globular cluster in Hercules; Btm.: M45, the Pleiades, an open cluster of stars in Taurus.

The Age of Clusters

A young galactic cluster will be more tightly packed with stars than an older cluster in which time has allowed the stars to drift apart. An H-R diagram plot of a young cluster has most stars following the main sequence, with the upper branch containing the blue stars curved away to the right. When older clusters are plotted, more stars further down the main sequence have moved toward the giant branch. Apparently, as stars age, the hot O and B blue giants are first to leave the main sequence to become red giants.

A plot of globular cluster stars finds the giant branch more fully developed. Stars down to +4 absolute magnitude have left the main sequence. In the age of the Galaxy, stars with the sun's magnitude have not as yet moved away. The abundance of variables in the clusters suggests an evolution beyond the giant stage. The H-R distribution of the stars shows a sharp curve away from the main sequence to the upper right of the giant branch. The stars continue across in the direction of the RR Lyrae variable stars of spectral type A and zero absolute magnitude.

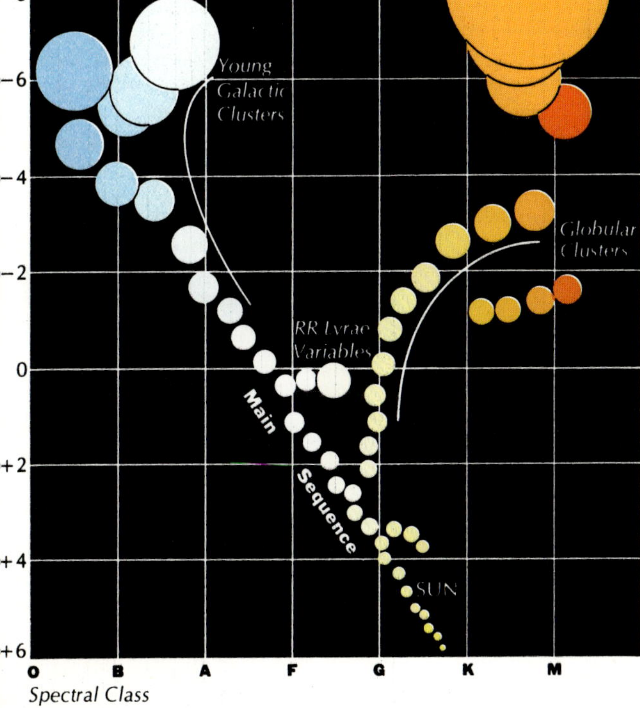

Spectral Class

Pulsating and Exploding Stars

The RR Lyrae variables found in the globular clusters represent one class of pulsating stars, but others with varying brightness were known long before the discovery of the cluster variables. *Mira*, a giant red star in Cetus, is an example of a *long period variable*, taking about 330 days to complete its period from maximum to minimum and back to maximum brightness again. Mira (''the Wonderful'') is too faint to be seen for about five months during minimum magnitude. At maximum, the star increases to third magnitude and is visible for about six months. This periodicity was recorded by the German astronomer Fabricius in 1596. In 1784, an Englishman, Goodricke, found the star Delta in the constellation Cepheus varying in brightness in a period of 5.36 days. Delta is the prototype of a class of variables known as *cepheid variables,* which have periods between one and 50 days. In 1912, the American astronomer Henrietta Leavitt discovered a relationship between the period of a cepheid and its luminosity. With luminosity known, absolute magnitudes are found to range from -1.5 to -5, which is bright enough to allow the stars to be seen at distances too great for a distance measurement by the parallax method. The pulsating cepheids seen beyond the Milky Way provide a yardstick to the distant galaxies.

There are stars called *planetary nebulae* with shells of expanding gases. In the telescope, these shells give the appearance of the disk of a planet. The gas envelope around the central star was emitted no more than a few thousand years ago. Prior to the planetary stage, the star may have been a red supergiant, which has a large cool envelope and a hot center.

Other stars called *novae* suddenly explode, brighten to about -9 absolute, and gradually fade to their pre-nova magnitude. A *supernova* outburst is even more dramatic with the star increasing in brightness to as much as -20 absolute magnitude. Several famous novae have been observed. In 1054, the Chinese recorded a supernova in Taurus that is now seen as a chaotic mass of gases called the Crab Nebula. (See p. 122.)

Star Dust and Gas Clouds

The *Great Nebula in Orion* is the best known example of irregular bright nebulosity. It can be seen with the naked eye as a hazy patch in the sword of the Hunter, south of the three stars marking the belt. In the telescope, the gases glow with a greenish, ethereal color. Centered in this luminous mass is the *Trapezium,* a cluster of hot, newborn stars responsible for the ionization that causes the nebula to glow. Other nebulae include the *North America Nebula* in Cygnus and the *Lagoon* and *Trifid* nebulae in Sagittarius. Near the bright nebulae are dark lanes and rifts, irregular in form and devoid of stars. These are the dark

Top lt.: NGC 7293 planetary nebula in Aquarius; the central star is ejecting shells of gas. Top rt.: M42, Orion Nebula; gases ionized by hot stars. Btm.: Pleiades, M45; starlight reflected by nebulosity surrounding the stars.

nebulae which appear to be connected and associated with the bright nebulosity. Type-O and -B stars ionize the atoms of the nebulae, which in turn emit radiation and glow by fluorescence. A nebula will remain dark in the absence of stars. Since bright nebulae shine by ionization, they are called *emission nebulae.*

Five stars in the Pleiades are surrounded by nebulosity that has the appearance of fleecy clouds with long filaments. These are *reflection nebulae,* which shine by starlight reflected from very small solid particles. Nebulae, which are generally restricted to the plane of the Milky Way, are associated with young stars of spectral class O and B that are still on the main sequence. Evidently, nebulae constitute the raw material from which the stars in the galaxy are born. In the Orion Nebula, stars are found surrounded by a primordial gas shell of hydrogen, attesting to their recent emergence.

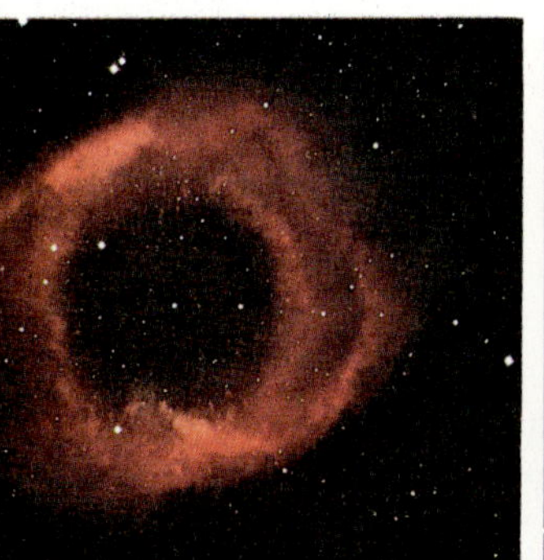

Stellar Evolution

Comparisons between recent and earlier photographs of the Orion Nebula show evidences of the condensation of stars. Dust and gases are associated with variable stars that exhibit rapid and irregular changes in brightness. These are believed to be young stars recently emerged from the emission nebula. A star is born when part of the nebula collapses, forming a central condensation that releases gravitational energy in the form of radiation. More hydrogen is added to the new body, or proto-star, which becomes opaque, preventing the loss of energy to radiation. Contraction causes the temperature to rise and density to increase until a balance is achieved and new hydrogen is no longer added to the star. The proto-star continues to contract, taking millions of years to reach the next evolutionary stage, which occurs when the temperature in the center reaches millions of degrees and thermonuclear reactions begin. The hydrogen in the star is converted to helium with a release of energy that radiates into space. Inside the star, two protons or nuclei of the hydrogen atom join to form a heavy isotope of hydrogen. Later, the heavy hydrogen joins with another proton to form a helium nucleus and a *photon* of radiation.

The amount of energy produced from mass is expressed in Einstein's famous formula, $E = mc^2$, where E, the energy, is equal in value to the converted mass, m, multiplied by c^2, the speed of light squared. (The hydrogen bomb exhibits a similar nuclear reaction, where very little mass produces a great deal of energy.) Now the star is said to be "hydrogen burning." The rate at which energy is produced depends upon how much hydrogen was present when the star was born. A giant blue star with a high temperature will consume more hydrogen in a shorter period of time than a smaller, cooler yellow star of the sun's proportions. The sun "burns" hydrogen at a faster rate than a small red dwarf star. All the stars on the main sequence are converting hydrogen to helium.

Creation of the Elements

Eventually the time will come when a star has converted all its available hydrogen fuel to helium. This occurs sooner in O- and B-type stars than in the sun and later in M-type red dwarfs. The blue giants are extravagant with their hydrogen supply and cannot remain on the main sequence for more than a few million years. The less massive red dwarfs are converting hydrogen more leisurely and reach ages of 15 billion years. The sun's lifetime on the main sequence is estimated to be about 10 billion years. Since the sun is believed to be 5 billion years old, it has already spent half of its time as a main sequence star.

    The H-R plots of clusters of various ages have shown that the more

luminous stars are first to leave the main sequence, with red giants appearing in their place. When the helium reaches about 12 percent of the total mass of the star, the core contracts, increasing the density, pressure, and temperature. The star's luminosity also increases, requiring a larger surface area to radiate energy into space. This is accomplished by a relatively rapid expansion to the red giant stage.

The red giants become the crucibles for the synthesis of the heavy elements. Now internal temperatures have reached 100 million degrees, which is hot enough to convert the helium core to carbon. In successive layers, the star burns carbon, helium, and hydrogen. The process continues as the carbon is synthesized to oxygen, neon, and magnesium, with each element created in its own concentric sphere at increasing temperatures. Finally the star produces an iron core which is the last and heaviest element to remain stable under temperatures of hundreds of millions of degrees. During this time of nucleosynthesis, the star moves on the H-R diagram horizontally to the left and back to the red giant stage several times, a course that is not fully understood. The abundance of RR Lyrae variable stars in the globular clusters indicates that the stars spend a time as short period variables after the giant stage.

In order to continue their evolution, stars must also lose excessive mass. Perhaps the star becomes a planetary nebula, casting off its outer layers. More violent explosions—as observed in novae and supernovae—complete the process. Heavier elements beyond iron (such as the radioactive elements) are created and released into space by the explosion. Later, these heavy elements mix with primordial hydrogen to form second generation stars similar to the sun. Heavy elements can only be produced within a star; therefore, the earth and other members of the solar system owe their existence to this evolutionary process.

M1, the Crab Nebula in Taurus,
was observed by the Chinese as a supernova
explosion in 1054 A.D.

121

White Dwarfs

Continuing the evolutionary process, the star moves inexorably toward the end. After shedding mass during the eruptive stage, the star must be reduced in mass to 1.2 solar masses or less to become a *white dwarf*. A white dwarf is a star that can no longer contract to produce radiant energy, since all thermonuclear reactions have already taken place during earlier stages. A thin, radiating outer layer provides the only remaining radiant energy. Eventually, this too will cease, creating a black dwarf as the final product of stellar evolution. The interior of a white dwarf is remarkable, for it is a star with most of the electrons stripped from the nuclei of the atoms: the electrons are free but squeezed closer together than in a neutral atom. This substance is called a *degenerate gas*, though it is much denser than any solid found on earth. A white dwarf is composed of *degenerate matter*. Contrary to the expected, a white dwarf star of one solar mass will have a smaller radius than another with only half the sun's mass. The size of the star decreases as its mass increases. If a white dwarf has a mass 1.2 times that of the sun, its radius must be zero. The largest white dwarfs are about twice the size of the earth with 0.2 solar mass. An increase in mass to that of the sun would reduce the white dwarf to the size of Mars.

Pulsars

In 1967, rapid radio signals with clock-like regularity were recorded by English astronomers. These signals, with periods of less than one second, came from stars that were named *pulsars*. The pulses indicate a rotating object that must be smaller than a white dwarf to be able to rotate so rapidly. The object responsible for the pulses is a strange *neutron star*, whose existence had been predicted some 40 years earlier.

A neutron star is denser than a white dwarf and only a few miles in diameter. In a neutron star, electrons have been forced into the nucleus of the atom, forming a neutron gas, while the outer layer of the star remains a rigid layer of neutrons. The most famous pulsar is found in the Crab Nebula in Taurus, where the Chinese observed a supernova in 1054 A.D. The high energy radiation of the Crab Nebula was difficult to explain prior to the discovery of the pulsar. Evidently, pulsars are following an alternate track after the nova stage. A star cannot become a white dwarf if it is unable to lose sufficient mass as a nova to bring it below 1.2 solar masses. Upon contraction, the excessive mass causes the star to shrink below the white dwarf stage to a diameter of about 6 miles. It has a thin gaseous radiating layer covering a solid shell, below which is a superfluid layer similar to liquid helium. The nature of the core is unknown, making a pulsar one of the mysterious discoveries of recent times.

A star in orbit with a black hole; gases from the star are attracted to the collapsar, resulting in x-ray radiation, which identifies the object.

Black Holes

A third possible terminal stage of a star is commonly referred to as a *black hole*. Cameron, one of the investigators of this phenomenon, has called such a star a *collapsar*. If a star has more than twice the mass of the sun before collapse, it will contract into a small object of such high mass that its enormous gravitational force will not allow energy—even light—to escape. Moreover, all energy from behind it and, indeed, all light falling upon it will be absorbed, making it not invisible—which would mean that we could see objects through it—but opaquely black. Outside, remnants of the star may surround the black hole like a shell of solid particles; inside, our earthbound concepts of space and time are so distorted as to have no meaning.

The possible existence of black holes was predicted almost 40 years ago. Recent observations have convinced astronomers that some x-ray sources detected by earth-orbiting satellites may be the mysterious collapsars. Stars have been observed to be revolving around invisible objects that emit x-rays and infrared radiation, and these objects behave like the predicted black holes.

Another object is Epsilon Aurigae, in the direction of Capella in Auriga. Epsilon is an *eclipsing binary*; the orbit seen edge-on from the earth, causing the stars to eclipse one another in a period of 27 years. The primary component is a giant yellow star; the secondary is invisible and radiating infrared rays. This secondary star is larger than the orbit of Saturn and may consist of the solid remnants of a star orbiting around a black hole. The primary star heats the particles around the collapsar, causing them to radiate in wavelengths beyond the visible range.

A Universe of Galaxies

The Galaxy

The *Galaxy* or *Milky Way* is a huge spiral wheel of stars about 30,000 parsecs or 100,000 light-years in diameter. Photographs of the Milky Way show intricate structure in the direction of Sagittarius. Here the stars are so numerous that their images overlap on the photographic plate. Star clouds abound as well as bright and dark, sinuous, meandering nebulae. Globular clusters surround the region like a halo suggesting the direction to the center of the Galaxy. Away from Sagittarius, the Milky Way appears striated with rifts of dark nebulosity that follow the galactic plane. Opposite Sagittarius, in the Orion region, the Milky Way, though less pronounced, is easily identified by the brightest constellations in the sky.

What appears from earth as the continuous band of the Milky Way is really three separate spiral arms of the Galaxy. The Sagittarius arm lies between the sun and the central region which cannot be observed photographically; the sun is located in the Orion arm, which contains the stars of Cygnus and Carina; beyond is the Perseus arm, about 2,000 parsecs from the sun. The Sagittarius arm is also about 2,000 parsecs from the sun, and the center of the Galaxy five times as far or 10,000 parsecs away. In the Orion arm, the sun is located near the inner edge, while the bright stars of Orion are found on a spur-shaped formation on the outer rim. The entire Galaxy rotates, at the sun's distance, at a speed of about 170 miles per second. At this rate 200 million years will be required for one rotation. Unlike a wheel, where the hub travels slower

than the rim, the stars closer to the center travel at a faster rate than the
sun and the stars in its vicinity, while those in the Perseus arm, in
accordance with Kepler's Third Law, are moving slower. However, the
stars from the center to about half the sun's distance rotate as though
they were part of a solid.

The spiral structure is more easily seen on photographs of more
distant galaxies. At best, only a portion of a few arms of the Galaxy can
be observed by optical and radio telescopes. Optically, the bright
spectral class-O and -B stars trace the arms. Radio telescopes detect
21-cm radio frequency radiation from the abundant neutral hydrogen
nebulosity. Radio waves can penetrate through the dust clouds and
radio telescopes can "see" deeper into the Galaxy than can optical
telescopes. Since the O- and B-type stars are associated with the
nebulae, astronomers using optical and radio telescopes together have
traced the arms and determined the structure of the Galaxy. Radio
telescopes have reached the nucleus, which appears to be a highly
concentrated ball of radiation, several parsecs in diameter, resembling a
gigantic star. Around the nucleus is a halo of stars, more concentrated
toward the center, then gradually thinning out at distances approaching
the outer arms. This halo contains the oldest known stars, including
globular clusters, RR Lyrae variable stars, and red giants. The spiral arms
follow the plane of the Galaxy and contain young stars imbedded in the
dust and gases of the nebulae. Gases are believed to be streaming into
the nucleus from intergalactic space and away from the center, into the
spiral arms to form future stars. The stars of today are destined to provide
heavy elements for the next generation to populate the Galaxy.

Satellites of the Milky Way

The nearest galaxies beyond the Milky Way are two irregular aggregations visible in the southern hemisphere. These are the *Magellanic Clouds,* named after the explorer who was first to circumnavigate the globe. The stars in the Clouds are main sequence stars surrounded by dust and gas similar to those found in the arms of the Milky Way. The galaxies are at a distance of about 50,000 parsecs or one and one-half times the diameter of the Milky Way. They are satellites of the Galaxy, held in place by the combined gravitational effect of all the stars. The distances to these galaxies were determined at the beginning of the century by Harvard astronomer Leavitt. She found *cepheid variable stars* in the Magellanic Clouds while photographing the southern sky at the Harvard University station in South Africa. The *period-luminosity relation* which was developed from these observations led to distance determinations. The total mass of the galaxies can be estimated once the distance and brightness is known. The Large Magellanic Cloud is equal to 20 billion suns in mass or about 10 percent of the Milky Way. The Small Magellanic Cloud has a mass equal to 2 billion suns. There are stars and gases forming bridges connecting the two clouds with each other as well as with the Milky Way.

Although the Large Cloud is classified as irregular, it does have structure. At each end of a pronounced bar of stars through the center, arms of stars are situated symmetrically, suggesting that the Large Cloud is a *barred spiral* similar to many found beyond the Milky Way. There are young luminous blue giant stars. Unlike the Milky Way, there are no large populations of old red giants and RR Lyrae stars in the Magellanic Clouds, which are therefore believed to be of later development than the Milky Way and to have evolved recently, near to and under the gravitational control of the Galaxy.

The Large and Small Magellanic Clouds, two
irregular satellite galaxies of the Milky Way; Opp.: NGC 205,
elliptical satellite of the Andromeda Galaxy

A Neighbor Galaxy

A large spiral galaxy can be viewed with the naked eye. Located in the constellation of Andromeda, the *Great Galaxy* measures one degree of arc (twice the angle made by the diameter of the full moon). This galaxy, called M31, is similar in structure to the Milky Way. In 1924, the American astronomer Hubble announced the discovery of variable stars in the *Andromeda Galaxy*, and by comparing the absolute and apparent magnitudes of the stars he estimated the distance to the galaxy. Since then, more accurate measurements place the galaxy at about 700 kiloparsecs or over 2 million light-years distant. This means that the light now received from the Great Galaxy in Andromeda left that star system when man emerged on the earth, more than 2 million years ago.

Through the telescope, the galaxy shines with a hazy glow, and long-exposure photographs are required to resolve individual stars and clusters. The central region contains a nucleus of red stars similar to the giants of the Milky Way. The Andromeda Galaxy is inclined about 15° to the earth, showing its opposite spiral arms winding around the nucleus several turns. Dust clouds line the inner edges of the arms. Luminous blue giants are strung out like beads in stellar associations. There are two visible satellite galaxies and possibly two more unseen but detected by radio telescopes. The two visible satellites are of the elliptical type and differ from the irregular Magellanic Clouds that accompany the Milky Way. These satellites contain old stars that are surrounded by globular clusters similar to the nucleus of a spiral galaxy.

Above: The nucleus of M31, the Andromeda Galaxy; Opp.: The galaxy in Triangulum, M33, a neighbor of the Milky Way.

The Local Group

In the universe, there is the tendency for objects to group together, beginning with planets in the solar system to star clusters and galaxies. The Milky Way and Andromeda Galaxy have satellite galaxies. Together, these two giant spirals are on opposite ends of an even larger aggregation of about 20 galaxies called the *Local Group*. The dimensions of the Local Group can be visualized by picturing the diameter of the Milky Way (which is 100,000 light-years across) as half the distance between the earth and moon. On this scale, the diameter of the Andromeda Galaxy would be equivalent to the moon's distance; the galaxy itself would be located at a point on the other side of the earth's orbit.

Most of the members of the Local Group are associated with the Milky Way or the Andromeda Galaxy. Exceptions are two small, irregular galaxies that are almost equidistant from these two gigantic spirals. Only one other spiral is found in the Local Group. This is M33 in Triangulum, which is near M31, along with six smaller, elliptical galaxies. Two recently discovered neighbors of the Milky Way are Maffei I and II, bringing the total number of galaxies associated with our Galaxy to nine. Motion accompanies gravitation, and all these galaxies are moving about a common center of mass located between the Andromeda Galaxy and our own. Measurement of this motion enables us to calculate the total mass of the system: the Local Group has a mass equivalent to that of 500 billion suns.

Various Kinds of Stellar Systems

Galaxies in the Local Group and beyond are classified according to their structure and the kinds of stars they contain. Hubble, who identified M31 as a spiral galaxy, devised a scheme for grouping the galaxies, but the system is too simple to include all the galaxies and should not be construed as suggesting a way in which galaxies evolve. The irregular galaxies along with other unusual types are not placed on the diagram. Hubble grouped the elliptical galaxies from a spherical E0 to E7 with greatest flattening. There are two general types of spiral galaxies, the normal spiral (S) and the barred spiral (SB). Starting with tightly wound spiral arms, SO, the classes divide into two branches according to the relative size of the nucleus and the development of the arms. A normal spiral such as M31 and the Galaxy have arms that begin at a spherical nucleus and unwind around the center. An Sa galaxy has a large nucleus and tightly wound arms. Barred spirals have a bar-shaped center with arms emerging from the extremities of the bar. Following the same classification as the normal spirals, an SBa has a large nucleus and tightly wound arms, an SBc a small center and loose, pronounced arms around the center. The elliptical galaxies, like the nuclei of the spirals, contain old stars that have already completed the hydrogen-burning stage of main sequence stars. Irregular galaxies and arms of spirals are composed of young stars surrounded by dust and gas. Irregular galaxies of young stars are extremely rare and may form when unusual concentrations of primordial matter occur in intergalactic space. Other galaxies are giant elliptical EO types that can be twice the mass of the Milky Way. Another unusual star system called a *Seyfert galaxy* has a bright central region and is believed to be related to distant objects called *quasars, or quasi-stellar radio sources.*

Quasars

Seyfert galaxies have extremely bright centers with spiral arms similar to those of other galaxies. If a Seyfert were located deep in space, only the bright nucleus would be visible. Similarities are found between objects without arms, the so-called quasars and the Seyfert galaxies. Quasars are believed to be the most distant objects observed. If this is so, they are brighter than −23 absolute magnitude, with diameters several light-years across—greater than the distance between the sun and the nearest stars. Seyfert galaxies and quasars exhibit similarities in visual, infrared, and radio investigations. Other explanations suggest that quasars are fragments of an original fireball explosion that created the universe. Another proposal holds that the quasars exploded out of the center of the Galaxy, eliminating the need to explain their enormous size and luminosity.

Top: Spiral galaxy, Pegasus; Whirlpool Galaxy, Canes Venatici; Mid.: Exploding galaxy, Ursa Major; Spiral galaxy, Sculptor; Btm.: Quasar 3C-295 in Bootes.

Clusters of Galaxies

At a distance of several diameters of the Local Group there are other clusters of galaxies. North of Spica in Virgo, there is a region called the Realm of the Galaxies. Here, vast assemblages are found containing thousands of galaxies in one of the largest clusters known. Its distance is about 40 million light-years. The number of galaxies contained may be numbered in the tens of thousands, with the greatest concentration toward the constellation Virgo; the number of galaxies tapers off as the Local Group is reached. From all appearances, this is the *local supergalaxy*, a galaxy of galaxies instead of stars. Our Galaxy, the Milky Way, is one in tens of thousands far out on the rim of a gigantic wheel of galaxies whose flat distribution indicates a rotation about the center. The local supergalaxy is only one of many, for investigations show others in the direction of Hydra and Pavo. Clusters of galaxies extend into space in all directions many billions of light-years. Thus, we can see that the universe is composed of hierarchies of increasing magnitude, in constant motion, in an unimaginably vast sea of space.

132

Above: Cluster of galaxies in Hercules;
various types of galaxies are seen,
including spiral galaxies in collision.

The Expanding Universe

Distances to M31 and other nearby galaxies can be measured by observing the cepheid stars and applying the period-luminosity relation. More distant galaxies require a statistical method. Assuming an average luminosity, a galaxy's brightness will diminish with distance. A galaxy twice as distant as another will be one-fourth as bright. This relation is called the inverse square law. Another method to determine distance employs the *Doppler effect*. The light from stars and galaxies is passed through a spectrograph which spreads the light into a spectrum of rainbow colors. Radiation from the interiors of stars is absorbed in their atmospheres, causing certain dark lines to appear in their spectra. If the star or galaxy is approaching the earth, these dark lines will be shifted toward the violet end of the spectrum; if the star is receding, the lines will appear shifted toward the red. The amount of shift will depend upon the speed of approach or recession.

When the spectra of the galaxies beyond the Local Group are examined, all of them show dark lines shifted to the red, indicating recession. The fainter and more distant galaxies show a greater red shift than those observed nearer the Local Group. There is a direct relationship between the rate of recession and distance, called *Hubble's law*. If the red shift is a Doppler effect, the universe is expanding, and galaxies are moving away from one another with increasing distances between the stellar systems, suggesting that at one time the universe was smaller, with galaxies closer together.

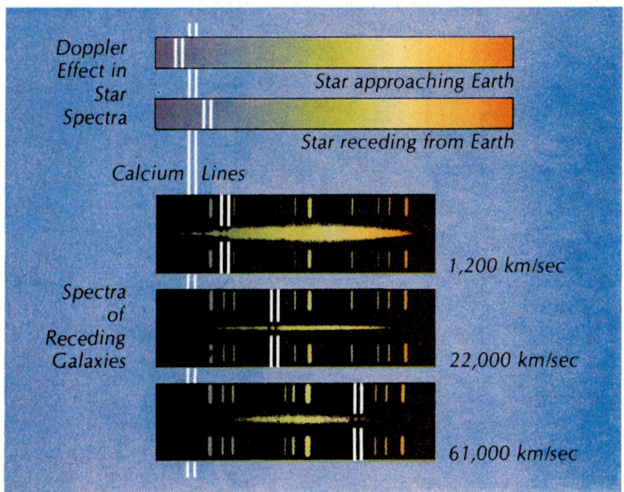

Continuous Creation

The "steady state" hypothesis of Bondi, Gold, and Hoyle, tries to circumvent the difficulty inherent in the assumption that the universe uses up all its substance, by introducing the concept of *continuous creation*. Here hydrogen atoms originate spontaneously between the galaxies, and this newly created hydrogen will condense into clouds which form the stars and galaxies. The galaxies still rush into limbo in an expanding universe but in the space between the galaxies, new stars and stellar systems are formed out of the newly created hydrogen. The number of galaxies per unit of volume remains constant, or in a "steady state." The universe has always been and always shall remain the same. The galaxies will differ, but the appearance of the universe will be unchanged. The galaxies appear to be moving away from each other at greater speed as the distance between them is increased. Yet, there is a theoretical limit to the cosmic horizon, for eventually the remotest galaxies will be receding at the speed of light. Beyond this boundary, the universe continues endlessly outside the field of view.

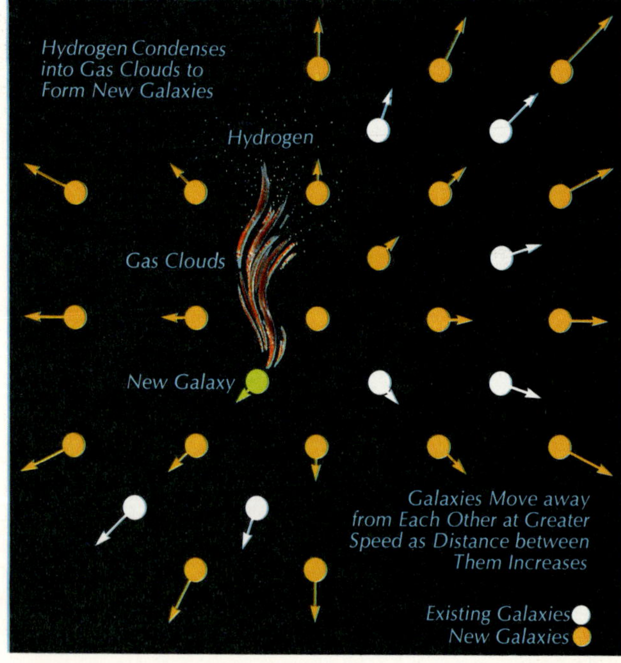

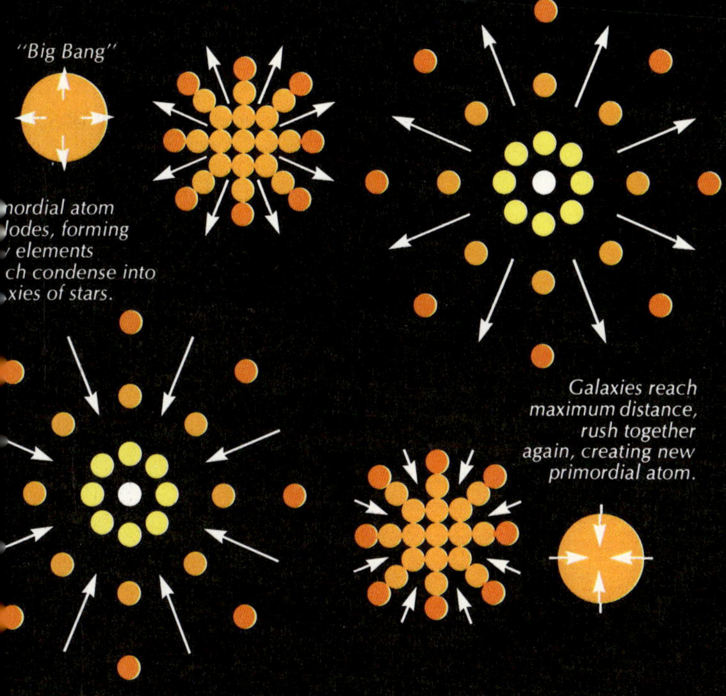

Evolution of the Universe

The origin of the universe as a primordial atom was proposed by G. Lemaître about 50 years ago. According to Lemaître's theory, the universe began as a small ball of energy that exploded, giving this hypothesis its popular name, "the big bang." Shortly, the elements were born in a rapidly expanding sphere forming into stars and galaxies. The universe continued to expand as the galaxies rushed into emptiness and will run down as all matter and energy permeate to an endless dimension. Will the expansion continue? Only if the primeval atom is rushing away at escape velocity. If not, as some theorize, the galaxies will reach a maximum distance and rush together again, forming a new atom of small size containing all the mass and energy of the present universe. Then the process will begin once more with an expansion to form new stars in a universe of another generation. This oscillating universe hypothesis removes one of the objections to the "big bang" theory. The primeval atom theory demands a beginning of time, but the matter and energy of the oscillating universe theory holds that matter and energy have always been, alternating from one state to the other.

135

Part 3•Satellite Exploration

Gemini 7, with Frank Borman and James Lovell, photographed from Gemini 6-A, with Walter Schirra and Tom Stafford, during first American rendezvous in space.

Earth Orbiters

During the International Geophysical Year (1957–1958), the nations of the world joined together in a cooperative scientific study of the planet earth. One of the features of the I.G.Y. was the decision by the United States and the Soviet Union to launch artificial satellites to investigate the earth from an orbit above the atmosphere. *Sputnik I*, a Russian satellite, was first in space on October 4, 1957. It was a sphere with a diameter of 58 centimeters (about 23 inches) and weighed almost 84 kilograms (about 184 pounds). Circling the earth every 90 minutes, Sputnik I transmitted signals from above the upper atmosphere (ionosphere) to stations of earth. Prior to Sputnik I, ionospheric investigations had been limited to high-altitude balloon flights and sounding rockets. Sputnik and the artificial satellites that followed ushered in a new age in the exploration of the earth and space. The first American satellite, called *Explorer I*, launched into orbit on January 31, 1958, discovered the inner *Van Allen radiation belt*, situated 1,800 miles above the earth.

Meteorology

Satellites have contributed much to meteorology and weather forecasting. The first, *Tiros I*, was sent into orbit on April 1, 1960. In an almost circular orbit at about 450 miles altitude, Tiros I photographed the earth's cloud cover with a vidicon TV camera and transmitted to stations on earth where pictures were reproduced from signals. For the first time, a global view of cloud structure gave advance warning of approaching hurricanes and typhoons. By August 1964, the much larger and more complex *Nimbus* was in orbit scanning with an infrared radiometer which instantly transmitted cloud cover information from the nightside of the earth.

Above: Nimbus III TV picture of clouds over southeastern U.S.; Opp.: left, Syncom at 22,300 miles from Earth; right, Relay Communications Satellite.

Communications

Another benefit has been the application of satellites designed to reflect and relay signals for communications. These fall into two categories: *passive,* which merely reflect earth-to-earth signals; and *active satellites,* which receive, amplify, and transmit radio signals. The first successful active-repeater satellite was *Telstar,* launched in July 1962, with a capacity of 60 simultaneous two-way telephone calls. Other active-repeaters are *Relay* and *Syncom;* the latter was placed in synchronous orbit of one revolution in 24 hours (earth's rotational period), keeping it fixed over the same part of the earth.

Astronomy and Earth Science

Some satellites observe the earth, the sun, and radiation in space. These are the *OAO* (orbiting astronomical observatories), using 36-inch telescopes to observe newborn stars and nebulae, and the *OSO* (orbiting solar observatory) to study the sun's x-ray emission. The *ERTS* (Earth Resources Technology Satellite) and *ATS* (Applications Technology Satellite) scan and photograph the physiographic and geographical features of the earth.

Future launchings include the *ATS-F* and an *SMS* (Synchronous Meteorological Satellite) to be placed into geosynchronous orbit for continuous weather study.

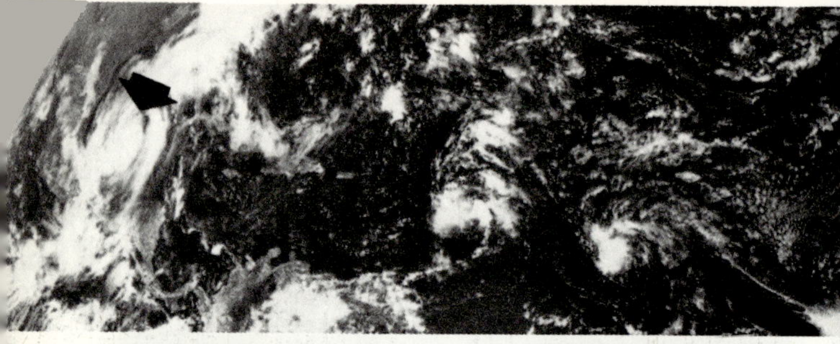

Top: New York-Philadelphia area from ERTS
Mid.: Storms tracked by satellite; Btm.: Cloud cover
over North and South America.

Exploration of the Moon

The Farside
On October 4, 1959, two years after the successful flight of Sputnik I, the Soviet Union launched the first space probe *(Luna II)* that photographed the hidden side of the moon. This farside is heavily cratered and unlike the nearside face, which is covered with flat lava plains. One large crater, named *Tsiolkovsky* after a Russian rocket pioneer, shows a dark flat floor with a central peak. In 1965, the Russians photographed the farside with the *Zond III* space probe and confirmed the lack of maria.

Lunar Closeups
During 1964 and 1965, three United States *Ranger* satellites photographed the moon and, shortly before impact, transmitted vidicon pictures which were the first high-resolution photographs showing details no larger than one-quarter mile and secondary craters formed by debris scattered during the formation of larger impact craters. The Ranger flights confirmed that the flat maria areas are acceptable as Apollo landing sites and gave strong support to the theory that most of the lunar craters were formed by impact.

Left: Lunar closeups prior to Ranger impact; Above: Ranger approaching the moon.

Soft Landings

The first soft landing was made in February 1966 by the Russian *Luna IX* space probe, proving that the lunar surface could support a manned expedition. The surface is rock-strewn and covered with a layer of fine dust. By June of the same year, the American *Surveyor I* made a soft landing on the floor of the Oceanus Procellarum with a TV camera that scanned a smooth rolling surface interrupted by craters from a few feet to a fraction of an inch across. Rocks are scattered about the moonscape partially submerged in a surface of fine granules. The camera showed the footpad of the spacecraft depressed about 2 inches in a soil-like layer that formed into clumps where disturbed by the landing.

About one year later, in April 1967, *Surveyor III* bounced to a landing on the side of a crater in the Oceanus Procellarum. Surveyor III carried a radio-controlled scoop which, on command from earth, dug a trench 6 inches deep, yielding material that has the consistency of damp soil as a result of vacuum cohesion. In November 1969, the Apollo 12 astronauts photographed this satellite and retrieved its camera.

Surveyor V landed in the Mare Tranquillitatis in September 1967 with an analyzer to determine the chemical nature of the moon. A box containing radioactive curium was lowered to the surface while alpha particles from the curium bombarded the surface and were scattered back again to the instrument, revealing a surface similar to volcanic basalt on earth. The TV camera showed small amounts of iron powder attached to a magnet fastened to one of the footpads.

Other successful Surveyor missions followed. *Surveyor VI* landed in Sinus Medii in November 1967 in preparation for the future Apollo flights. (The ill-fated Apollo 13 mission was slated to land here.) In January 1968, *Surveyor VII* was placed on the highlands near the crater Tycho and picked up laser beams transmitted from earth stations, studied the soil with an alpha-scattering instrument, and scooped up the surface with a sampler.

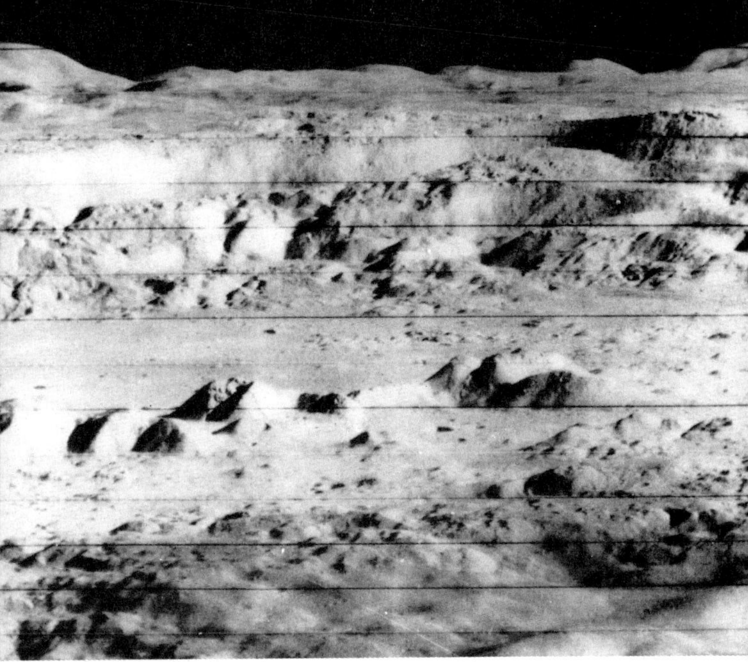

Orbiting the Moon

While the Surveyors were investigating selected surface areas, *Lunar Orbiter* spacecraft were obtaining closeup views of the entire moon. In August 1966, *Orbiter I* transmitted a spectacular photograph of a vast panorama of the moon's limb with the distant crescent of earth above the lunar horizon. Medium resolution pictures of possible Apollo sites showed the maria regions to be pockmarked by myriad small craters, contrary to the smooth appearance shown by earth-based telescopes.

In November 1966, *Orbiter II* continued the task of photographing the equatorial region in search of landing sites. A high-resolution photograph of the crater Copernicus became the "picture of the century." Orbiter II found the site where Ranger VIII had crashed on the Mare Tranquillitatis. By February 1967, *Orbiter III* had photographed the landing area of Surveyor I. Finally, detailed photographs of the farside were made with wide-angle and telephoto lenses.

Orbiter IV photographed the Orientale Basin, a bull's-eye formation probably created by impact; *Orbiter V* was the first to photograph an almost full earth. By the completion of the Orbiter series, virtually the entire moon had been photographed. New craters were named, the farside was charted, and landing sites were surveyed for the ultimate mission—*Apollo*.

143

Planetary Probes

Missions to Mars

The first closeup photographs of Mars were received in July 1965 during the Mariner IV flyby mission. Aboard the satellite, a vidicon image tube signal was stored on magnetic tape as digits, transmitted to earth and reconstructed into a picture by computer. This technique was employed in later Mariner missions which returned photographs showing craters similar to those of the moon.

By 1969, the Mariner VI and VII flyby missions revealed surface features unknown on the earth and moon. These include cratered terrains previously observed with Mariner IV; chaotic terrains of irregular structure with ridges and depressions suggesting erosion; and feature-less terrains of circular "deserts" without structure. What appear as "canals" through earth-based telescopes are alignments of craters and irregular dark regions.

In 1971, the Mariner IX orbital mission photographed 85 percent of the planet. At the same time, Russia's Mars II orbital satellite ejected a capsule to the surface, but radio transmission ceased almost immediately after descent.

The Mariner IX flight confirmed the irregularity of the Martian surface—craters with wave-like surface texture suggesting "sand" dunes. A meandering "arroyo" 355 miles long resembles a dried river bed although the existence of former rivers is unconfirmed. A volcano called Nix Olympica is 335 miles in diameter and towers 15 miles above the surrounding plain. A "grand canyon" called Mariner Valley is almost 4 miles deep and stretches across the planet for more than 2,000 miles, reaching a width of 150 miles.

Pioneer to Jupiter

Jupiter was reached by satellite for the first time on December 4, 1973, when *Pioneer X,* after almost two years in flight, passed within 78,000 miles of the cloud-covered planet. In addition to the exploration of Jupiter, Pioneer X studied the asteroid belts with an asteroid-meteoroid detector and found travel through this region of thousands of minor planets not hazardous to the spacecraft. Other instruments on board designed to study the planet included magnetometers, photometers, a Geiger-tube telescope, a cosmic-ray telescope, and a plasma analyser.

The probe began to measure a magnetic field and a magnetosphere containing high-energy electrons and protons surrounding the planet. With an energy 250,000 times greater than the earth's magnetic field, Jupiter's magnetism deflects particles from the sun as far as 4 million miles from the planet.

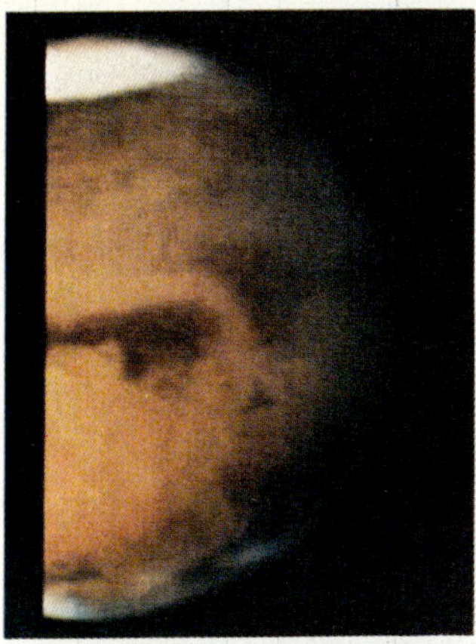

Left: Mariner flights to Mars provided closeup views of surface features; Below: In December 1974, Pioneer XI revealed a lack of atmospheric bands in the polar region of Jupiter.

Pioneer X found that the planet radiates two and one-half times the energy received from the sun in the form of infrared radiation. If Jupiter had been larger, it would have become an incandescent star, since the planet is chemically similar to the sun but is not sufficiently massive to generate the internal temperature and pressure required for the nuclear processes found in the sun and stars.

145

To the Inner Solar System

The first attempt to explore Venus took place on February 4, 1961, when the Russian *Sputnik VII* was launched. The mission ended in failure, but in 1965, *Venera III* became the first object to impact the planet. *Venera IV* (June 1967) attempted to land an instrumented capsule.

In the United States, exploration of Venus began in July 1962 with *Mariner I*, which failed to reach the planet. *Mariner II*, launched about one month later, came within 21,000 miles of Venus. As the Mariner passed the planet, an infrared radiometer scanned from the nightside to the dayside, measuring brightness and temperatures. Surface temperatures, deduced from the readings telemetered back to earth, indicated a hot (800°F), dry planet incapable of supporting life.

The *Mariner V* mission of 1967 found the atmospheric pressure on the surface of Venus to be about 100 earth atmospheres—equal to the water pressure at the bottom of the ocean. This enormous pressure is

147

*Opp. top: Jupiter, televised by Pioneer X, December 10, 1973; Opp. btm.: Artist's view of Pioneer X over Great Red Spot
Above: Venus, televised by Mariner X, February 1974*

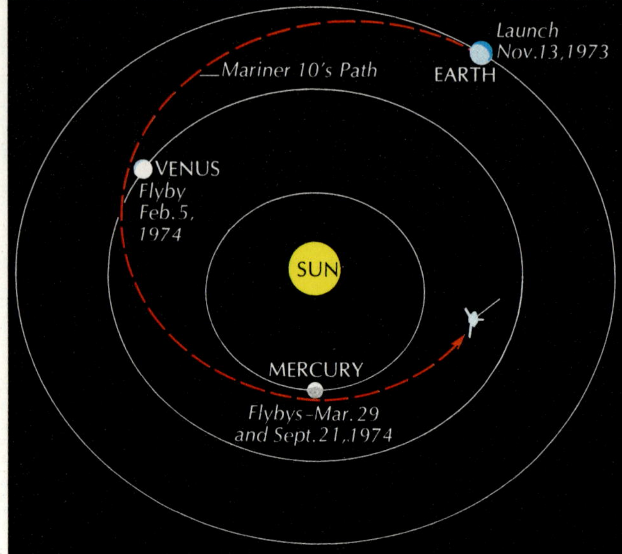

believed to have destroyed the Venera IV before it reached the surface of the planet. Later, Soviet missions of *Venera V* and *VI* in 1969 attempted soft landings with parachutes. These satellites came near the surface before radio contact was lost.

In 1973, space scientists took advantage of the alignment of Venus and Mercury to photograph both planets on one mission. By February 10, 1974, *Mariner X* had photographed the Venus cloud cover in ultraviolet light. In the equatorial region, the direct rays of the sun cause vertical circulation resulting in a huge ring structure in the clouds. Time-lapse photographs show a longitudinal circulation of the upper atmosphere at 250 miles per hour. Circumequatorial belts extend from the subsolar ring across the face of the planet while spiral streaks from the middle latitudes arc toward the equatorial belts. In the polar region the clouds form a polar ring of excess condensation.

By the end of March 1974, Mariner X arrived for its rendezvous with Mercury. Pictures were returned showing the planet to be cratered like the moon. Bright streaks radiate from large craters. Flat maria ringed with mountain chains are lacking, although the appearance of Mercury suggests bombardment by meteoric blocks from space. The planet has a weak magnetic field and an atmosphere of helium, both of which may **148** be the result of the impact of high-energy particles from the sun.

Top: Mercury from Mariner X in September 1974 from 47,000 miles; Btm. lt.: Mariner X ; Btm. rt.: Cratered surface of Mercury resembles the highlands of the moon.

Men in Space

Orbital Flights

The events that reached culmination with a manned landing on the moon began on April 12, 1961, when Yuri Gagarin in the Soviet craft *Vostok 1* made the first flight and circled the earth in one hour and 48 minutes. In the following month on May 5, 1961, Alan Shepard became the first American in space with a 16-minute sub-orbital flight aboard the *Mercury 3,* named "Freedom 7." On July 21, 1961, Virgil Grissom duplicated the feat aboard the *Mercury 4,* called "Liberty Bell," which sank during recovery operations. The next spectacular space event was provided by the Russian, Gherman Titov, aboard the *Vostok 2.* He made 16 revolutions and spent more than 24 hours in space. John Glenn became the first American in orbit on February 20, 1962, with three revolutions and a total flight time of four hours and 55 minutes in *Mercury 6.* The mission was repeated by Scott Carpenter aboard *Mercury 7* on May 24, 1962. In August, *Vostok 3* and *Vostok 4* with Adrian Nikolayev and Pavel Popovich made the first group flight within 3 miles of each other. In October 1962, Walter Schirra flew the *Mercury 8* for six revolutions. The program ended with Gordon Cooper in *Mercury 9* with a flight lasting 34 hours and 20 minutes.

Multiple Missions

The next two years were dominated by the *Vostok 5* and *Vostok 6* group flight in June 1963 (Valentina Tereshkova in Vostok 6 becoming the first woman in space); the first three-man *Voskhod 1* mission with Vladimir Komarov, Konstantin Feoktistov, and Boris Yegorov; and the March 1965 flight of *Voskhod 2* with Pavel Belyayev and Aleksei Leonov, who made the first EVA (extra-vehicular activity) with a 10-minute space walk.

From March 1965 to November 1966, the American Gemini program dominated manned space exploration. Virgil Grissom and John Young flew three revolutions in *Gemini 3* and became the first American "space twins." In June 1965, *Gemini 4* carried James McDivitt and Edward White for 62 revolutions with White performing the first American space walk for 21 minutes. Gordon Cooper and Charles Conrad made the first long space flight which lasted more than one week (190 hours and 56 minutes) in August 1965 (*Gemini 5*). But this record was broken on December 4, 1965, when Frank Borman and James Lovell in *Gemini 7* made 206 revolutions before splashdown (330 hours and 35 minutes). Meanwhile *Gemini 6-A* was launched on December 15, 1965, with Walter Schirra and Thomas Stafford to rendezvous with Gemini 7. The two spacecraft came within one foot of each other.

The remainder of the Gemini missions were spent practicing rendezvous, docking, and EVA. It was necessary to perfect these maneuvers before attempting a landing on the moon. On March 16, 1966, Neil Armstrong and David Scott made the first docking to the Agena target. Thomas Stafford and Eugene Cernan followed in June 1966 with 72 hours of rendezvous and EVA. Michael Collins and John Young docked with the Agena in July 1966. In September 1966, Charles Conrad and Richard Gordon continued rendezvous and docking. James Lovell and Edwin Aldrin completed the series aboard *Gemini 12* in November 1966.

Top: Edward White performing an EVA, photographed by James McDivitt aboard Gemini 4; Btm.: Agena docking target, photographed from Gemini 12 by James Lovell and Edwin Aldrin.

Steps to the Moon

The Russian cosmonauts continued their earth-orbiting flights with the new Soyuz series. *Soyuz 1*, the heaviest manned craft, was launched with Vladimir Komarov in April 1967. He was killed on re-entry after 17 revolutions. Earlier, in January 1967, the first American Apollo crew, Virgil Grissom, Edward White, and Roger Chaffee lost their lives in a fire while testing the Command Module on the giant Saturn V booster on the launching pad.

The first manned Apollo flight took place on October 11, 1968, when Walter Schirra, Donn Eisele, and Walter Cunningham tested the spacecraft in earth orbit for more than 11 days. The mission was successful, and on December 21, 1968, Frank Borman, James Lovell, and William Anders made their historic *Apollo 8* voyage to the moon, made 10 revolutions, and returned safely to earth. With *Apollo 9* in earth orbit, James McDivitt, David Scott, and Russell Schweikart docked the LEM (Lunar Excursion Module) and practiced EVA with a self-contained life-support system. Docking and rendezvous around the moon was accomplished in May 1969 by Thomas Stafford, Eugene Cernan, and John Young aboard *Apollo 10*.

Now the stage was set for the big moment—July 16, 1969—when Neil Armstrong, Edwin Aldrin, and Michael Collins were to start out for a lunar landing. The LEM, with Armstrong and Aldrin aboard, separated from the CM (Command Module) piloted by Collins. The LEM named Eagle landed in the Sea of Tranquillity on July 20, 1969. Six and one-half hours later, Armstrong stepped on the moon followed by Aldrin. The astronauts examined the surface, collected rocks and soil, and set up various instruments for scientific investigation. The following day, they rejoined Collins in the CM and started home for a successful splashdown on July 24. Man had reached the moon eight years after his initial orbital flight around the earth.

Opp. top: Lunar surface from Lunar Module window; Opp. btm.: Aldrin at base of ladder of Apollo 11 Lunar Module, photographed by Armstrong.

Later Apollo missions followed. In November 1969, Charles Conrad, Richard Gordon, and Alan Bean reached the moon in *Apollo 12.* Conrad and Bean brought the LEM to the Oceanus Procellarum near the Surveyor III and retrieved its camera. *Apollo 13* with James Lovell, Fred Haise, and John Swigert experienced an oxygen explosion on the way to the moon, resulting in a massive power failure. The astronauts returned unharmed. In January 1971, *Apollo 14* landed Alan Shepard and Edgar Mitchell in the Fra Mauro region, while Stuart Roosa remained in lunar orbit. For the first time, equipment and supplies were carried on a two-wheeled transporter. *Apollo 15* was at the Hadley-Apennine region in July 1971, David Scott and James Irwin using the Rover vehicle for the first time, Al Worden piloting the CM in orbit. Nuclear seismometers recorded signals 60 miles into the lunar interior. In April 1972 *Apollo 16* astronauts John Young and Charles Duke became the ninth and tenth Americans to walk on the moon while Thomas Mattingly orbited above. The LEM landed in the Descartes highland plateau. Investigations were made with a seismometer, magnetometer, and a cosmic-ray detector. The most productive mission was the *Apollo 17* flight to the Taurus-Littrow region in December 1972: Eugene Cernan and Harrison Schmitt explored the surface while Ronald Evans remained in orbit. The Apollo missions show the lunar crust to be deficient in iron and rich in calcium, aluminum, and titanium. The moon is older than was previously believed and different from the earth in composition.

Top lt.: Apollo 15 Command and Service Module in lunar orbit; Top rt.: Apollo 15 astronaut James Irwin with Lunar Roving Vehicle.

Beyond Apollo

The first *SKYLAB* was placed into earth orbit in May 1973. These space stations consist of an *orbital workshop* (a converted Saturn IV-B booster); *airlock module; multiple docking adapter;* Apollo *service* and *command module;* and an Apollo telescope mount with eight telescopes to observe in the x-ray and ultraviolet wavelengths as well as the visible spectrum. By February 1974, three SKYLAB crews had completed their missions, with the last crew setting a record of 84 days in orbit.

In 1975, the United States and Soviet Union plan to link an Apollo and Soyuz for the first international space mission. This is a step toward future cooperative missions in earth orbit and to the planets.

By 1978, the *Space Shuttle* should be operational. As large as a jet liner, the Space Shuttle will carry satellites to and from orbit and will transfer crews in space and *fly* back to a designated airport on a 78-foot delta wing. Space shuttle may be the prototype for future commercial flights into space. Space travel for everyone seems remote today but fewer than 70 years separate the first airplane flight by the Wright brothers from the first landing on the moon.

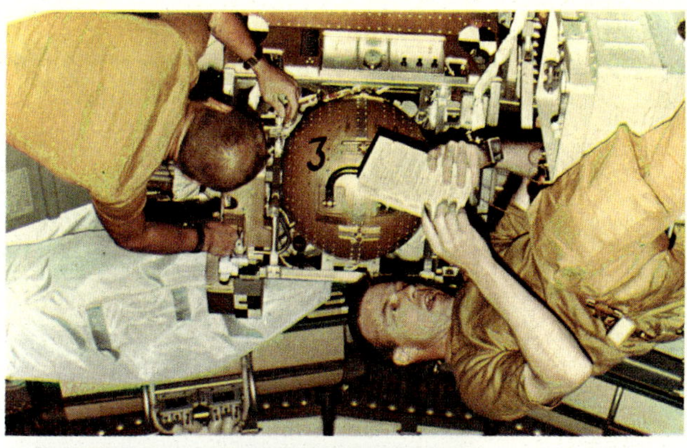

Above: Skylab 1 crew members Charles Conrad and Joseph Kerwin aboard the Skylab Workshop; Left: Skylab space station, taken from Skylab 2 Command Module.

Index

Add to your
KNOWLEDGE
THROUGH
COLOR

(All Books $1.45 Each)
(Where marked • $1.95 Each)

Listed on the opposite page are the currently available titles in this paperback series. To add to your KNOWLEDGE THROUGH COLOR library, simply list the books you want and mail to:

BANTAM BOOKS, INC.
Dept. KTC-1
666 Fifth Avenue
New York, N.Y. 10019

Add 25¢ to your order to cover postage and handling. Please send a check or money order, since we cannot be responsible for orders containing cash.

A BANTAM CATALOG IS AVAILABLE UPON REQUEST.
Just send your name and address and 10¢ (to help defray postage and handling costs) to: Catalog Department, Bantam Books, Inc., 414 East Golf Road, Des Plaines, Illinois 60016.